Metro Ethernet Services for LTE Backhaul

For a complete listing of titles in the
Artech House Mobile Communications Series,
turn to the back of this book.

Metro Ethernet Services for LTE Backhaul

Roman Krzanowski

ARTECH HOUSE
BOSTON | LONDON
artechhouse.com

Library of Congress Cataloging-in-Publication Data
A catalog record for this book is available from the U.S. Library of Congress.

British Library Cataloguing in Publication Data
A catalogue record for this book is available from the British Library.

Cover design by Vicki Kane

ISBN 13: 978-1-60807-685-7

685 Canton Street
Norwood, MA 02062

10 9 8 7 6 5 4 3 2 1

To Eliza and Jacob

Contents

Acknowledgments

No book is created in a vacuum. Several people contributed to this book directly by providing comments to the text or indirectly by discussing the specific topics or problems covered in the book. I have to thank Raghu for his general remarks on the concept of the book itself, Tom for initial encouragement and QoS insights, Brendan for his support of the project and insightful comments, Tony for his important input on several aspects of the book (in particular transport and protocol stacks), Mark from Artech House for the support of this project and an unwavering belief in its success, the anonymous reviewer for the devastating critique of the first version of the book, and Samantha for her guidance during the publication process. I also would like to thank the colleagues with whom I have worked over several years and who provided a unique working environment—Marje, Frank, Ron, Bob, Steve, Matt, Larry, and many others. Of course, any mistakes or inaccuracies in the book are solely of my doing and my responsibility.

1

Introduction

Radio has no future.

—Lord Kelvin, 1897 [1].

1.1 Why This Book

This is book is intended to serve as a guide to designing mobile Ethernet backhauling (MEBH) services in a metropolitan area using carrier Ethernet (CE) architecture. The components of the CE architecture are defined in many standards published by the Metro Ethernet Forum (MEF), the International Telecommunication Union (ITU), the Institute of Electronic and Electrical Engineers (IEEE), and the Internet Engineering Task Force (IETF) organizations. It is relatively easy to understand the specific concepts behind the elements of the CE service, such as, for example, quality of service (QoS), Ethernet virtual connection (EVC), user network interface (UNI), performance monitoring, and service operation, administration, and maintenance (SOAM) functions [2]. It is more difficult to put together out of these many elements available in the CE Service toolbox [3] a coherent working design for the specific service. The CE service architecture with its many components is like many jigsaw puzzles mixed together. As with all jigsaw puzzles, not all pieces fit well together. To assemble a puzzle, one has to know which pieces go with which one, what the rules are for combining the pieces, and what the final picture looks like.

No single guide shows how all CE service components with definitions dispersed in various standards and publications come together and interact and what are the interdependencies and consequences of (and reasons for) selecting specific components for the service. In several cases, MEF documents such as MEF 6.1, 10.2, 26, 23, ITU Standards, and IEEE [4] discuss the interdependencies of different aspects of the service. Yet, the format of these documents, demanding that they use the succinct language of requirements, limits the options that can be presented and the way they can be presented. Standards state the most general, or abstract, technical solutions. Specific choices for a specific design for the specific service are very particular, and the reasons for making this or that choice in this or that case are not in any publicly disclosed document; these come mostly from experience, not to say from trial and error (but that's what experience is after all!). Thus, the one who designs an MEBH service, while an expert in wireless technology, SONET/SDH transport, MPLS architectures, or other networking designs, may be unfamiliar with the CE services and thus may look puzzled at a list of MEF documents augmented by ITU, IETF, and IEEE specifications—not knowing where to start, where to end, and what to do in the middle of the design.

Obviously, this book will not give answers to all possible questions about the designing the MEBH service, and many, many details of the design will be left out. Nevertheless what the author hopes for is that this book will show how to construct the basic structure of the service using elements specified in standards identifying the dependencies and consequences of some of the choices that have to be made while designing the service.

Thus, the restated objective of this book is to first introduce the necessary components of the carrier Ethernet service and then show how the components can be assembled together to craft the MEBH service solution.

And the last word: this is not a book on how to engineer the MEBH service in every detail, down to specific configuration parameters, or configuration scripts, or parameters values; this also is not a guide to the MEF or ITU standards. This book describes show to create a framework for designing the MEBH service in metro. The framework is, by virtue of being a framework, high level, yet without the framework outlined in this book, the detailed design of the service will not have a firm basis, an organizing principle, or the necessary completeness. As well, while care is taken to define clearly the concepts discussed in the book, the book is not an introduction to fundamentals of the networking, and the author assumes on the part of the reader certain basic level of understating of the networking technologies.

The MEBH service is a service based on Ethernet technology standards designed to carry or transport traffic between cell sites and mobile transport switching office (MTSO) locations [5]. In the architecture of wireless networks, MEBH denotes the first segment of the network connecting the cell site facilities with the MTSO equipment. Figure 1.1 presents a high-level view of wireless networks and the location of the MEBH network segment in the metro Ethernet network. In most cases, MEBH is restricted to the metro area, usually understood as an area covered within a radius of no more than 120 miles. The

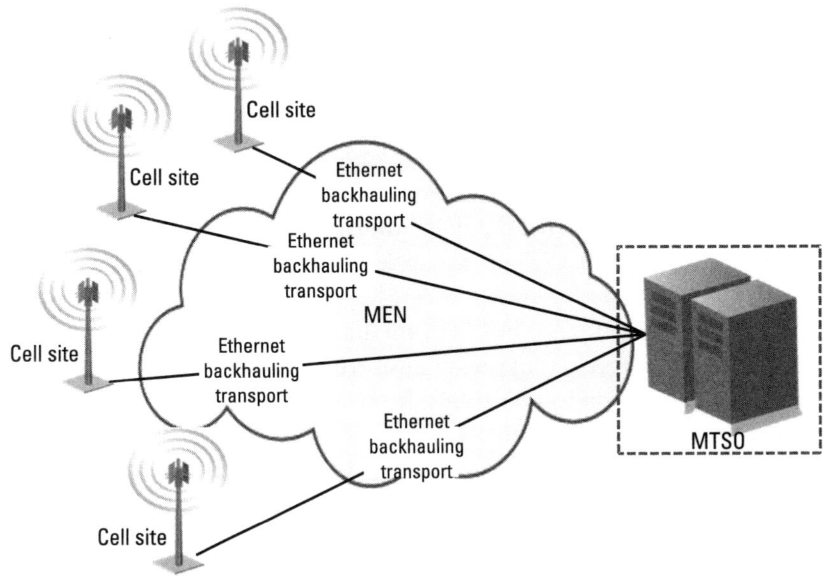

Figure 1.1 High-level view of MEBH service in MEN.

network providing the Ethernet service in the metro area is usually referred to as the metro Ethernet network (MEN). Of course, such a definition is valid for urban areas. In rural areas, MEBH may be extended over larger distances. In addition, the definition of the service does not exclude multiprovider networks in the metro area; quite often, in some locations the MEBH service is offered by two or more provider, as a single provider does not cover the entire service area.

Wireless backhaul may be differently understood by people depending on their scope of responsibilities. Some may extend mobile backhaul to include the radio equipment at the cell site as well as the equipment at the MTSO location. The definition of the MEBH service is fairly loose in specifying the exact demarcation points. However, the discussion in this book is limited to the segment of the wireline network between interfaces facing the cell site on one end and the MTSO on the other end and remaining under the management of the MEBH provider.

The radius of the MEBH service is dictated by the wireless technology requirements that set certain strict performance objectives for the traffic transport between the cell site and the MTSO equipment. These performance objectives would be impossible to support over large areas without a loss of the quality of the wireless service.

Why is MEBH such a hot topic? Several years ago, wireless carriers realized that their main service would evolve from carrying voice to carrying data. Handheld devices would become terminals for voice, video, and text, supporting the full range of Internet services, evolving into full-fledged multifunction data terminals demanding high bandwidth communication channels. At the same time, the number of different applications that could be implemented on the handheld devices would grow beyond the wildest projections. This evolution of cellular traffic would require explosive growth in wireless network capacity as well as in all the segments of the transport networks carrying the wireless traffic [6]. That was the reality to come.

In these early days, backhauling from the cell sites to the MTSO was accomplished by using T1 [7] or similar technologies. This architecture is presented in Figure 1.2.

T1 had its boom days, but with the limited capacity, cost, and maintenance problems, this technology was not suited to support the future growth in capacity foreseen by wireless providers. In the second half of the first decade of the 2000s, the diagram presented in Figure 1.3 spelling doom to this method of carrying traffic was circulated on all conference presentations and copied in a multitude of industry magazines. The diagram showed that with the T1 transport technology and the explosive growth in wireless traffic, the cost of providing backhauling (BH) transport with T1s would outstrip the revenues from the new data services.

A new method for transporting traffic from cell sites to MTSO locations needed to be found: a method mature in the market, more efficient, more reliable, and cheaper. Although with T1s wireless providers were able to carry several megabits of traffic from a single cell site to an MTSO, marketing predictions professed the need for 20, 50, and 100 Mbps or more. And this was only the beginning, we were told. With this scenario T1s were clearly out of the game. Of course, nobody of sane mind believed in these marketing curves predicting an exponential growth. But, surprisingly, this prediction, contrary to many other futuristic claims made in the past on a variety of topics, did become reality [8].

The choice of technology for MEBH was Ethernet or carrier Ethernet (CE), meaning the Ethernet network technology deployed for carrier-grade

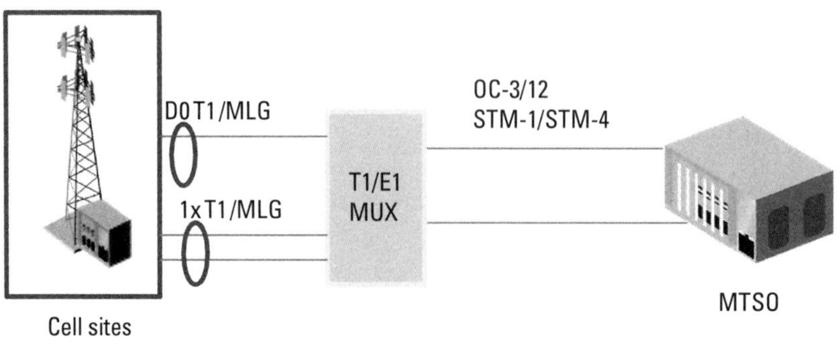

Figure 1.2 T1-based backhauling design.

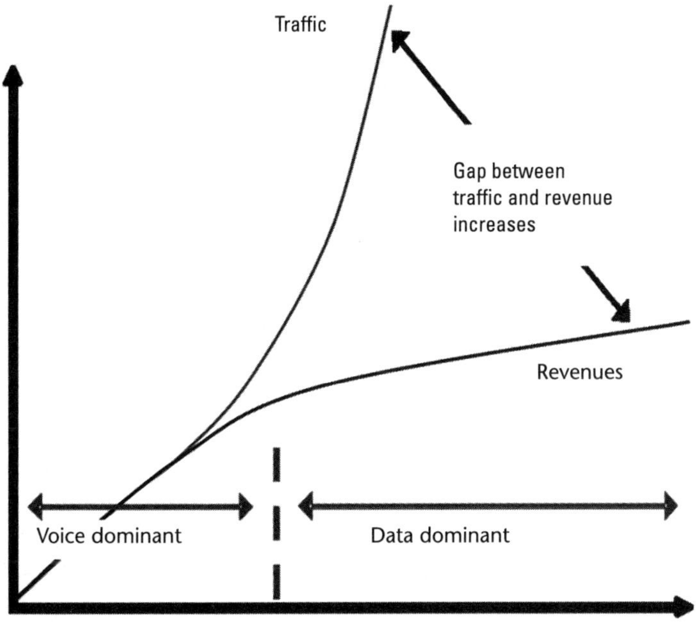

Figure 1.3 The "doom" scenario for wireless evolution (circa 2007–2008) [9].

networking services. Ethernet services have been maturing since the 1990s (Ethernet is last-century technology!) evolving from intra-office (enterprise) technology into carrier grade networks. Ethernet services were cheaper, more reliable (than traditional T1s) if correctly engineered, and seemingly ubiquitous in metro areas. The problem was that although network providers had extensive experience using Ethernet as a carrier technology, for mobile providers Ethernet was a new game. Yet the idea was born.

The past few years (2010–2013) have witnessed explosive growth in Ethernet backhauling services beyond even the most radical predictions. Looking toward the future, these predictions still defy any previous estimates [10]. As of writing of this book (mid-2012), some analysts predict that the combined market for new wireless gadgets and applications ranging from all kinds of smartphones, tablets, and ebook readers to a plethora of online applications such as Internet radio, video, TV, and many others probably yet unknown will increase the demand for backhaul ten times by 2016 [11]. Thus, there is no question that in the coming years the need for fast, cheap bandwidth will only grow at a staggering pace. The pressure for services from wireless providers will most probably force an evolution of the current carrier Ethernet MEBH designs to support per-usage billing, ubiquitous bandwidth on demand, asymmetric EVC profiles, a more flexible and reliable transport layer, reliability beyond "six nines,"[1] and almost lossless transport, to mention just a few of the possible changes to the way the current MEBH service operates. With all these new features implemented, we will witness a revolution in networking technology leading to ubiquitous wireless access, the Panopticon beyond the wildest dreams of its originator.

For now, it seems that Ethernet technology did fulfill its promise of being a ubiquitous, relatively cheap, and flexible transport media, and, if managed wisely, Ethernet will continue to be the technology of choice for wireless backhaul into the foreseeable feature. The question the industry is facing now is not whether to use Ethernet, but how to use it in the best way possible. Of course, this is a sign of some degree of maturity. An impediment in this process is the relatively limited understanding of Ethernet networking and its standards.

Hence, here are the purpose and objectives of this book restated. The book will explain the features of Ethernet technology and related service standards. The book will focus on services, their specifications, and key technology concepts that will allow those with some networking experience to navigate through the Ethernet technology maze. As we said, facing the universe of terms and concepts defining the Ethernet networking may be a daunting experience.

This book is written with the objective of easing these learning pains and making CE technology less obscure. The book is based mostly on Metro Ethernet Forum (MEF) and ITU service concepts. However, this book is not an MEF or ITU guide, nor is it intended as a review or a commentary on Ethernet technology standards published by these or other standard bodies. Thus, topics will not be treated with all their nuances and details. The purpose of the book is to demonstrate that by using the existing conceptual tools one can design an end-to-end carrier grade MEBH service framework and that one can take the concept of the service and build from it the service architecture. The author hopes that the book will cover all of the relevant aspects of CE technology for backhauling wireless traffic in a manner that allows the attentive reader to understand the CE specifications and service definitions, and construct requirements for the service and eventually a complete high-level design solution.

1.2 MEN and MEBH Service

We begin with two reference diagrams explaining the key concepts behind Ethernet services. Figure 1.4 presents the elements of Ethernet service. The customer premises equipment (CPE) [12]. is connected to the MEN service via the user network interface (UNI). UNIs delimit the Ethernet virtual connection (EVC) that spans the metro network. UNIs demark the service boundaries (i.e., locations where the service provided by the MEN provider ends and

1. "Six nines" is a measure of the reliability of a service.

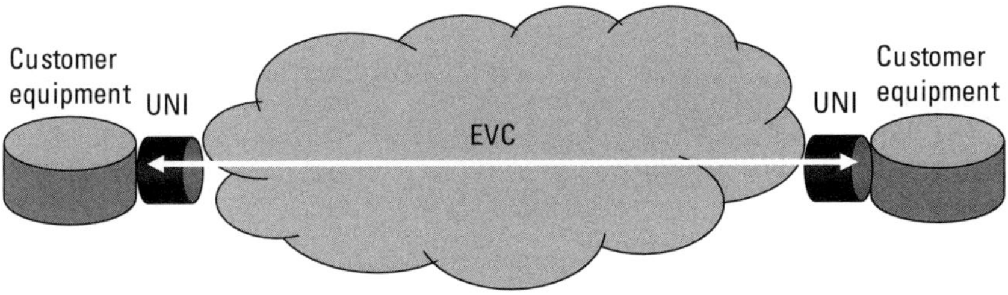

Figure 1.4 Elements of the MEN service.

begins). Each MEN Ethernet service can be represented in the canonical form as depicted in Figure 1.4.

A high-level view of the MEBH service over the MEN is depicted in Figure 1.5. The MEBH service transports the traffic between the mobile network radio access network base station (RAN BS) site, which is usually a cell site, and the mobile network radio access network network controller (RAN NC) site, which is usually an MTSO. The segment of the network between these sites is realized using Ethernet technology. One can easily map the elements of the MEN service from Figure 1.4 to the elements of the MEBH service in Figure 1.5 The customer sites—RAN BS and RAN NC locations—are represented here by CPE. The MEBH service demark points are UNIs, and the mobile backhaul is realized by EVC or the Ethernet virtual circuit.

In Table 1.1, the Ethernet MEN service and MEBH concepts are set side by side so one can see how they are related.

Focusing on the MEN for MEBH service is a bit restrictive, or, more precisely, it gives a somewhat incomplete view of the scope and nature of MEBH

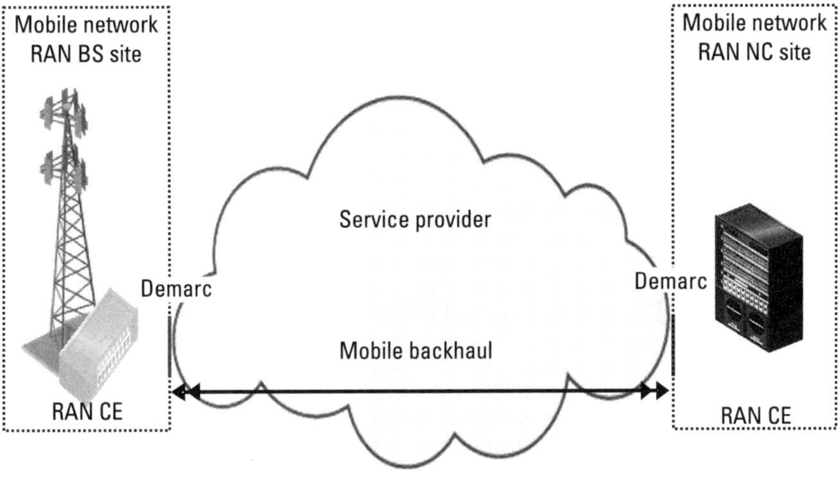

Figure 1.5 Ethernet backhauling high-level diagram.

Table 1.1
Comparison of MEBH and MEN Service Elements

Mobile Network	Metro Ethernet Network	Function
Mobile network RAN BS	Customer premises equipment	Equipment at the customer site
Mobile network RAN NC	Customer premises equipment	Equipment at the customer site
Service demark	UNIs	Interfaces to the network
Mobile backhaul	Metro Ethernet network	Network transport between customer sites
MEBH transport circuit	EVC	Logical connection between UNIs

services. The MEBH service in many cases extends beyond the single MEN, if by MEN we understand the network in the metro area under the management of a single authority. Quite often the transport from the cell site to the MTSO must be accomplished with two or, in some cases, more than two, service providers. This situation arises if a single provider does not have the reach (service coverage) guaranteeing service to the cell site or to the MTSO locations. An example of MEBH composed of two segments is presented in Figure 1.6.

In the multiprovider MEBH service, the overall framework of the MEBH service with an EVC and UNIs bounding the EVC has been retained. Two additional constructs are added: an operator virtual connection (OVC) and an external network-to-network interface (ENNI). The ENNI is the interface between two networks. The OVC extends from one UNI to the ENNI or between two ENNIs. The OVC is a segment of the EVC between one UNI and the ENNI and within a single service provider network. Thus, a multiprovider (or multisegment) MEBH service is composed of at least two OVCs and one ENNI, in addition to the EVC and UNIs. One can imagine the situation when

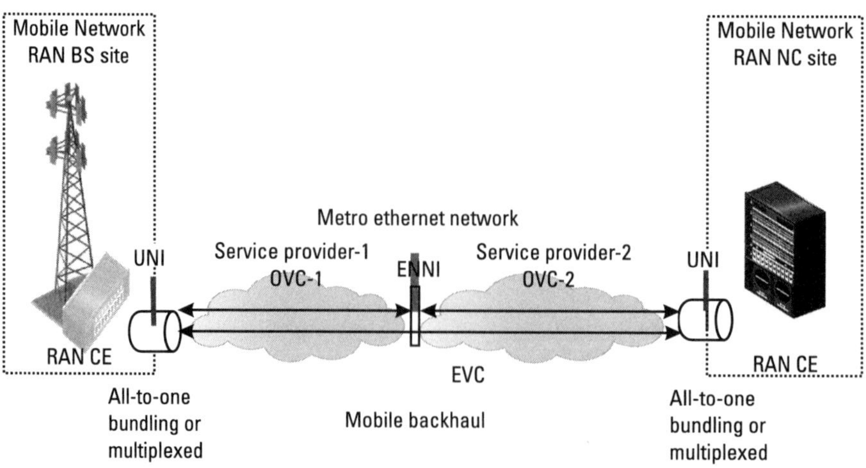

Figure 1.6 MEBH service with two service providers.

more than two service providers are supporting the EVC, as illustrated in Figure 1.7. Although such architectures are certainly possible, they are quite infrequent. Stitching the EVC from two or more OVCs adds to the complexity of the design and service, and such situations should be avoided.

To account for the multiple providers, we could then extend Table 1.1 and add required elements, as illustrated in Table 1.2.

The customer in principle should not be aware of the OVCs; the end-to-end service design is the responsibility of one of the providers.

The entire book will be about how to map or use the services available in the MEN technology and represented in Figure 1.4 to architecture Ethernet backhaul service as defined in Figure 1.5 (and Figure 1.1), what aspects of MEN technology are important to the MEBH service, what these technologies mean in the context of this service, what choices one may make within the MEN technology, and what the consequences of those choices are. Thus, in essence the book is about how to translate the technology depicted in Figure 1.4 into the services in Figure 1.5.

1.3 LTE Architecture Overview

The technologies that will give wireless carriers the capacity needed to support all new data applications are referred to under the umbrella term of long-term evolution (LTE) architecture. Although we cannot do justice in this book to the

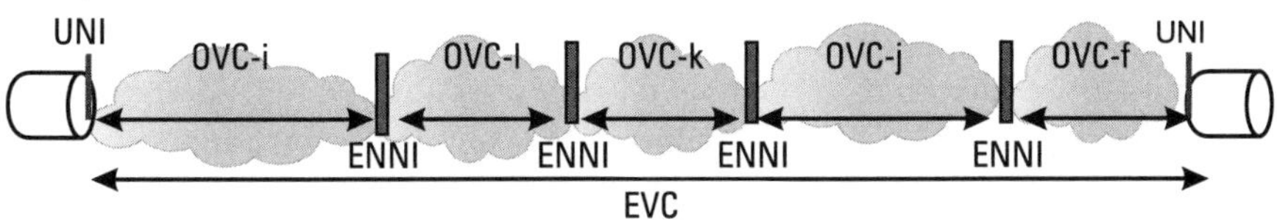

Figure 1.7 MEBH service with multiple service providers.

Table 1.2
Comparison of MEBH and MEN Service Elements for Multiple Service Providers Case

Mobile Network	Metro Ethernet Network	Generic Function
Mobile network RAN BS	Customer premises equipment	Equipment at the customer site
Mobile network RAN NC	Customer premises equipment	Equipment at the customer site
Service demark	UNIs	Interfaces to the network
Mobile backhaul	Metro Ethernet network	Network transport between customer sites
MEBH transport circuit	EVC	Logical connection between UNIs; may be composed of several OVCs, connected at ENNIs

complexity of LTE, a brief overview is needed, at least introducing some key terms that may be encountered in further discussion in this book or in other documents related to the MEBH of LTE. The presented material draws mostly on LTE resources available on the 3GPP Organization's [13] web site.

The newest mobile networks are marketed as third-generation (3G) radio technologies, while LTE is marketed as 4G technology. LTE promises >100 Mbps downlink and >50 Mbps uplink speeds with up to 1 Gbps for advanced LTE, reduced latency (RTT < 10 ms on user plane), compatibility with the existing spectrum, compatibility with older wireless technologies, and reduced per-bit cost. Existing 3GPP (GSM and WCDMA/HSPA) and 3GPP2 (CDMA2000 1xRTT, EV-DO) systems can be integrated in the evolved system through standardized interfaces. Figure 1.8 illustrates the evolutionary path toward 4G wireless technologies [14].

The LTE architecture includes the evolved UMTS terrestrial radio access network (E-UTRAN) segment and the nonradio aspects under the term system architecture evolution (SAE). The entire system composed of LTE and SAE is called the evolved packet system (EPS). At a high level, the network is composed of the core network (CN) and the evolved packet core (EPC). The CN is responsible for the overall control and establishment of the bearers. The LTE (LTE/SAE) architecture is presented in Figure 1.9.

The MEBH service is part of the E-UTRAN segment of the LTE architecture. The service is designed to support X2 and S1 interfaces between eNBs and S-GWs. The eNB or eNodeB are evolved base stations. X2 is the interface between the eNBs used primarily for handoffs, and S1 is the interface toward the core between eNBs and the SAE gateway, serving gateway (S-GW) [16], or the user plane entity (UPS). The MEBH network is not, in most designs, connected directly to eNBs and S-GWs, but its logical function is to support traffic flow for control and data planes between the cells site equipment and the S-GW and mobility management entity (MME) complex.

Figure 1.10 depicts the LTE interfaces, S1 and X2. The LTE user plane is effectively reduced to two nodes—the LTE base station evolved node B (eNodeB) and the SAE gateway. The control functions are located in the MME

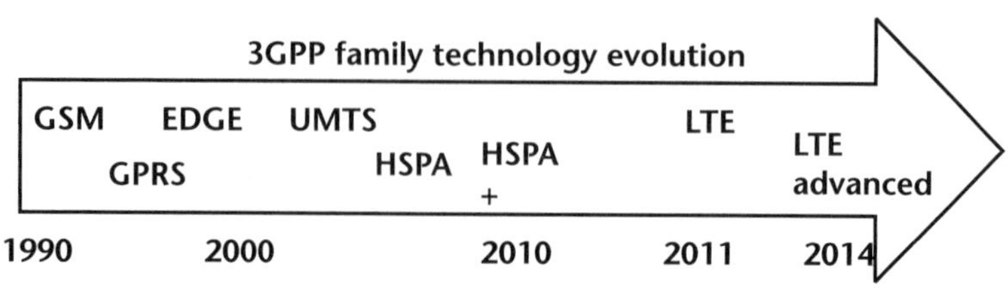

Figure 1.8 Evolution of wireless networks toward LTE and 4G [15].

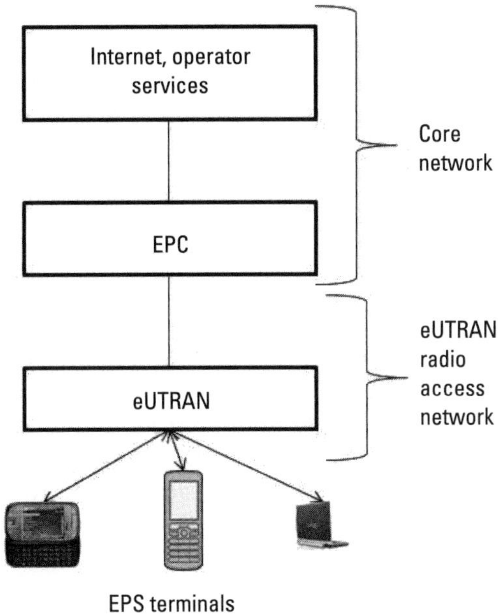

Figure 1.9 LTE architecture.

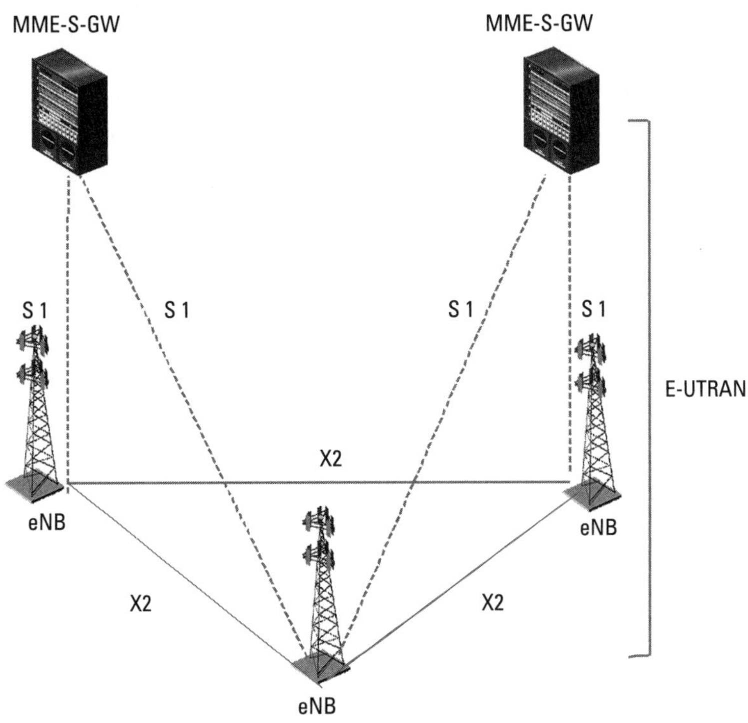

Figure 1.10 High-level LTE interfaces [17].

node, separate from the gateway (S-GW). A slightly more detailed model of the LTE is presented in Figure 1.11.

Figure 1.11 shows that the S1 interface carries two channels—the control S1-MME and the user S1-U. The X2 interface provides intercell connectivity. This architecture will have a significant impact on the design of the MEBH, as some designs will establish an EVC per an interface, rather than to lump two interfaces into a single EVC. These design choices will be discussed in the second part of the book.

1.4 The Road to Ethernet

In the early days of MEBH, the transition from the architecture in Figure 1.2 to the architecture in Figure 1.12 was not as easy, obvious, or straightforward as it may seem now. Everyone would agree that the architecture to use is the one in Figure 1.12. The T1 was an outdated, not scalable, outage-prone technology, so it had to go. The question was how to migrate to Ethernet with the minimum disruption to service [19].

With no Ethernet interfaces in the wireless equipment, providers, if they wanted to benefit from the Ethernet technology, had three options. They could use existing T1 facilities and interface with the Ethernet network using the gateway routers converting T1 to Ethernet-compatible protocols for the data plane and use the T1s and T1 infrastructure for the control plane (option A in Figure 1.13). They could build into the wireless equipment an Ethernet interface, retain the T1 interfaces, and dispense with the gateway routers (option B in Figure 1.13); this option would retain some of the T1 infrastructure as well. They could also eliminate the T1 infrastructure altogether and use the Ethernet transport for the use and control planes (option C in Figure 1.13), if they had enough confidence in the Ethernet technology and gateway routers. In these

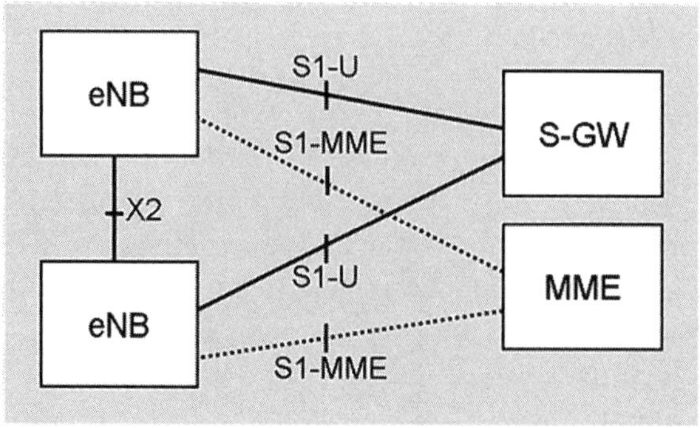

Figure 1.11 LTE X2 and S1 interfaces [18].

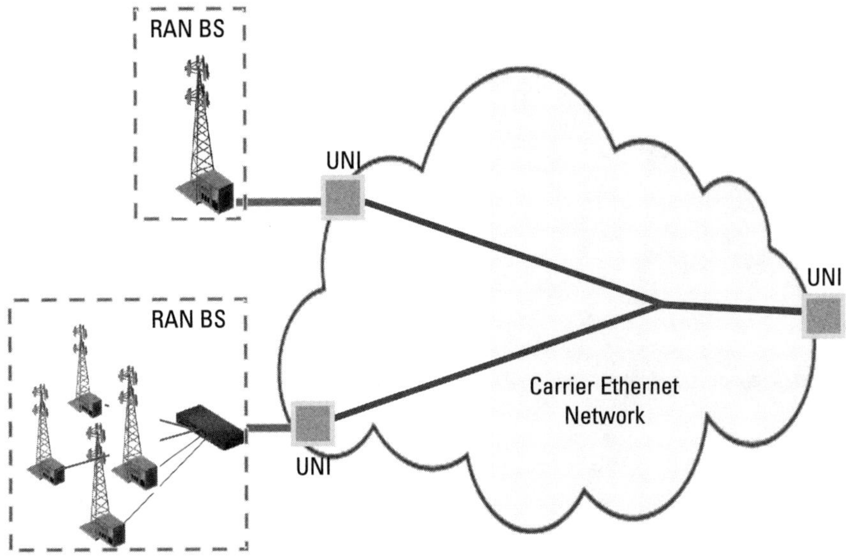

Figure 1.12 Target MEBH architecture.

early days, all of these options had been deployed, and some are still supported in cases where the move to the Ethernet architecture and the LTE technology has not been completed.

However, eventually, with the LTE architecture all wireless carriers will have to end up with fully compatible Ethernet interfaces in the cell site and MTSO equipment that will allow them to dispose of any of the T1 technologies and embrace fully Ethernet (option D in Figure 1.13). That is where we are roughly today (2013) for most major carriers.

1.5 Outline of the Book

The book is about how to construct the MEBH service using the components of the Ethernet technology. It takes the view of the MEBH that the customer of such a service would have. The customer and the service provider look at the same service from different angles; the customer defines the service specifications, and he is, in principle, not overly interested in the design details as long as he gets what he wants. He has this famous "black box" view of the MEBH service (we will see that the customer cannot afford to have such a limited view of services he purchases). The MEBH service provider thinks and talks network,

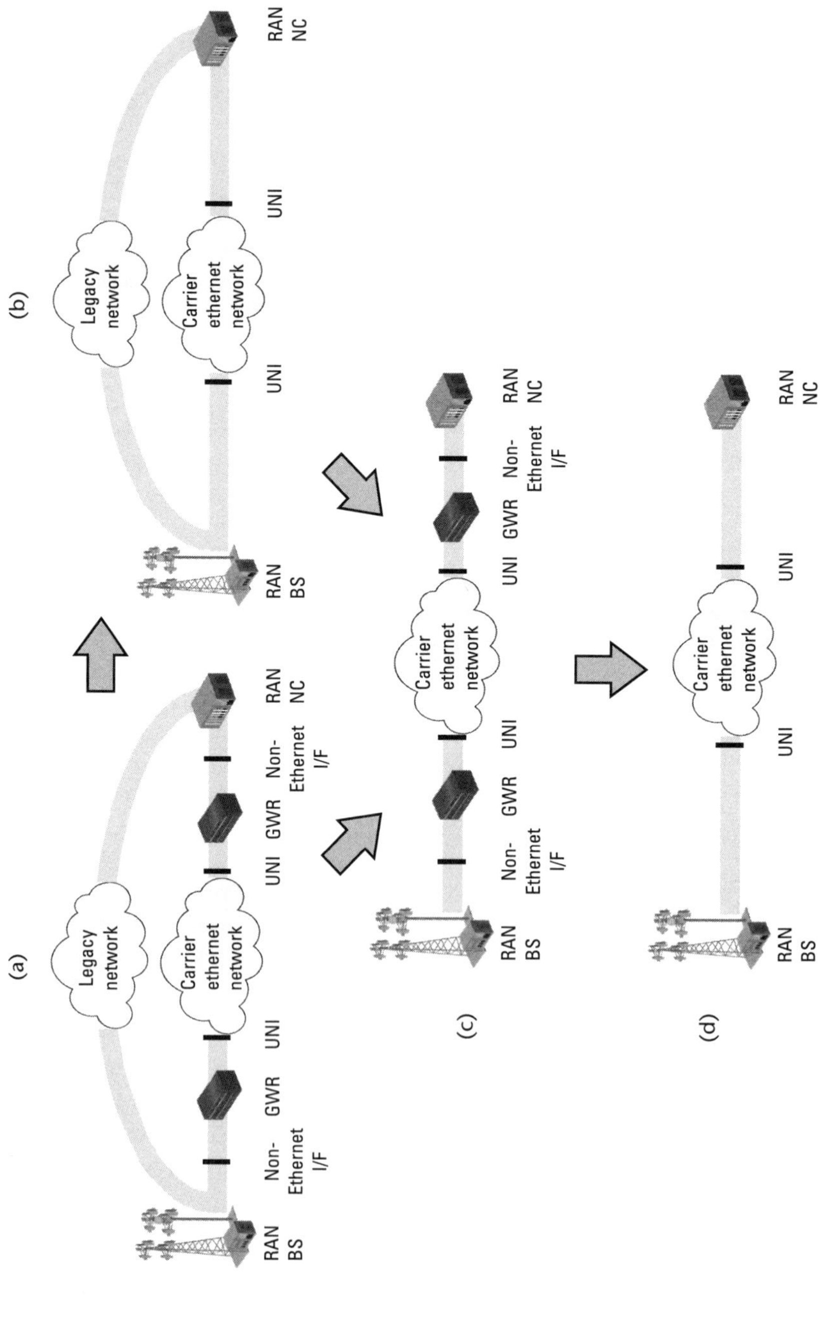

Figure 1.13 MEBH evolution from T1 to Ethernet.

and she must comprehend the customer's service requirements in the context of her technology. These two perspectives have to meet. The best results are obtained when both understand each other's lingua and objectives—they sing from the same key. And this is what this book is trying to facilitate.

Thus, the next section will discuss the two views of the same service in more detail—the view of the provider and the view of the customer. The section after that presents in some detail Ethernet technology. The discussion is segmented into the service functional groups that reflect the customer's view of the Ethernet service. The section following this discussion reviews the Ethernet technology in each of the service functional groups from the perspective of the MEBH service. The book is completed with the section presenting the exemplar use case of the MEBH service. The use case contains the specification of the service architecture using both the customer's view through the service functional groups and provider's as reflected in Ethernet service objects attributes. Putting both perspectives of the same service side by side shows how these two views come together, somewhat proving the closure on the whole book.

A word of caution; the book will not talk about any of the mobile technologies such as GSM [20], WCDMA [21], CDMA2000 [22], WiMAX 802.16e [23], 3G [24], 4G [25], and 3GPP or LTE [26]. The discussion will be mostly limited to single MEN with interfaces between RAN BSs and the RAN NC. However, the book will briefly present basic architectures involving multiple service providers, ENNI attributes, and the concept of the OVC, fundamental elements of the multiprovider architecture [27].

On a personal note: with respect to the newest technology I prefer as a reference to use resources available on the web, conference proceedings, and conference presentations simply because these are more nimble channels of communication than books and paper publications are. Those contain in my view a lot of operational experience and ideas that would never make into a book or would make it too late for the fast-evolving technology that is networking. There is a place, and a role, for the printed word, no doubt. And there is a place, and a role, for electronic information. The web allows quick access to most recent experience and ideas, which is not easily achievable with the book format. This book tries to be somewhere in the middle. That is why in this book there is a balance of references from books and resources available on the web. Some may see it as a weakness; I see it as a sign of times. But, as with everything, in selecting resources one must use good judgment. So, no Twitter references are found here.

Endnotes

[1] Lord Kelvin, c. 1897. Quoted in Cerf, C., and Navansky V., *The Experts Speak*, New York: Pantheon Books, 1984.

[2] One may also encounter an acronym EOAM denoting SOAM for Ethernet services.

[3] For some, no doubt, the set of networking standards is more akin to the den than to the toolbox.

[4] See the Appendix: Ethernet Standards for a list of Ethernet-related standards from several standard bodies.

[5] "Mobile or wireless backhaul is the portion of a wireless network that connects information traveling from a wireless tower to a mobile switching center." Byme, D., "What Is Mobile Backhaul?" http://www.ehow.com/facts_7406132_mobile-backhaul_.html. Accessed March 23, 2012.

[6] It may come as a surprise to many that the biggest segment of the wireless services infrastructure is composed of the wireline networks.

[7] "The T1 is what telephone companies have traditionally used to transport digitized telephone conversations between central offices. The bandwidth of a T1 is commonly known to be 1.544 Mbps. This represents the maximum bit carrying ability of a T1. The overhead necessary to frame a T1 is 8 Kbps. Therefore, the total usable bandwidth is 1.536 Mbps, or the equivalent of 24 DS-0 channels. A single DS-0 has a bandwidth of 64 Kbps and is designed to carry a digitized telephone call. Today, T1 technology is being used in private and public networks to carry both voice and data traffic." *T1 Networking Made Easy*, Pulse Technical Handbook Series, Pulse, Inc., 2004, http://www.pulsewan.com/data101/pdfs/t1basics.pdf. Accessed May 19, 2012.

[8] See, for example, Metcalfe, R., *Internet Collapses and Other InfoWorld Punditry*, Foster City, CA: IDG Books Worldwide, 2000.

[9] "Carrier Ethernet for Mobile Backhauling," MEF, http://metroethernetforum.org/PDF_Documents/posterofmbhfinal.pdf. Accessed April 2008.

[10] Cisco Visual Networking Index, "Global Mobile Data Traffic Forecast Update, 2010–2015," http://www.cisco.com/en/US/solutions/collateral/ns341/ns525/ns537/ns705/ns827/white_paper_c11-520862.html. Accessed November 10, 2011.

[11] Golstein, P., "Study: U.S. Mobile Backhaul Demand to Grow Nearly 10x by 2016," *Free Wireless Daily Newsletter*, http://www.fiercewireless.com/story/study-us-mobile-backhaul-demand-grow-nearly-10x-2016/2012-03-13. Accessed March 22, 2012.

[12] The CPE acronym is sometimes interpreted as customer provided equipment.

[13] 3GPP, http://www.3gpp.org/About-3GPP. Accessed August 12, 2012.

[14] Artza Networks, "LTE Tutorial," http://www.artizanetworks.com/lte_tut_eut_arc.html. Accessed May 15, 2012.

[15] 4G America. "LTE: Long Term Evolution," http://www.4gamericas.org/index.cfm?fuseaction=page&sectionid=249. Accessed May 15, 2012.

[16] Ibid.

[17] Hamza, A. "A LTE Architecture—A Tutorial," Network Systems Laboratory, Simon Fraser University, http://nsl.cs.sfu.ca/wiki/upload/1/10/LTE.pdf. Accessed May 15, 2012. Ericson, A. "Long Term Evolution, An Interlocution," *Event Helix*, http://www.eventhelix.com/lte/tutorial/lte_overview.pdf. Accessed May 15, 2012. Fritze, G. "SAE—

The Core Network for LTE," Ericsson Austria GmbH, 2008, 3G4G Wireless Resource Center, www.3g4g.co.uk/Lte/SAE_Pres_0804.Ericsson.pdf.

[18] Olsson, J., "Mobile Backhaul," MEF, 2009, http://metroethernetforum.org/PPT_Documents/MEF_CE_MBH_Arch_pub-Jonathan_Olsson-TNMO.ppt, Accessed May 22, 2012.

[19] Ibid.

[20] GSM (Global System for Mobile Communications, originally Groupe Spécial Mobile) is a standard set developed by the European Telecommunications Standards Institute (ETSI) to describe technologies for second-generation (2G) digital cellular networks.

[21] Wideband CDMA is a third-generation (3G) wireless standard using one 5-MHz channel for both voice and data, initially offering data speeds up to 384 Kbps.

[22] Enhancements to the CDMA cellular technology that increase the number of users on a voice circuit and provides high-speed packet data. In 2000, CDMA2000 was the first 3G technology to be deployed as part of the ITU's IMT-2000 framework.

[23] WiMAX (Worldwide Interoperability for Microwave Access) is a communication technology for wirelessly delivering high-speed Internet service to large geographic areas. WiMAX is based upon IEEE Std 802.16e-2005

[24] 3G or third-generation mobile telecommunications is a generation of standards for mobile phones and mobile telecommunication services fulfilling the International Mobile Telecommunications-2000 (IMT-2000) specifications by the International Telecommunication Union.

[25] 4G is the fourth generation of cellular mobile communications standards. It is a successor of the third-generation (3G) standards.

[26] 3GPP Long Term Evolution, referred to as LTE and marketed as 4G LTE, is a standard for wireless communication of high-speed data for mobile phones and data terminals.

[27] A good general introduction to LTE technology and mobile backhaul (coming from Nokia Simens Networks) can be found in Metsala, E., and J. Salelin, *Mobile Backhaul*, Chichester, UK: John Wiley and Sons, Ltd., 2012.

2

Constructing MEBH Service

What the hell it is good for?

—R. Lloyd, IBM, 1968, regarding the microprocessor [1].

2.1 Design Process—Perspectives

The wireless provider designing MEBH service encounters a swarm of Ethernet technology concepts that may seem bewildering and disconnected at first. Figure 2.1 gives a (very limited) glimpse of the Ethernet technology concept space. To know where to begin the design of the service and how to progress requires a good deal of experience and familiarity with both the technology as implemented and underlying standards.

It is hard to make sense of which concepts are fundamental, how concepts are related and what are dependencies between them [2]. One may of course start at any point, but not all starting points are created equal. Some concepts are more fundamental and constraining on the design than others. Thus, the Big Questions are: Which of these concepts are critical? Where do we start? How do we progress?

There are different ways to specify the Ethernet service, or organize and structure the service concepts. All depends on the view of the service one has. The service provider will present its services in a set of attributes of service objects—EVC and UNI, as listed in Table 2.1.

The provider's Ethernet service specifications are well packaged (thanks to Metro Ethernet Forum work) as the UNI profile, EVC profile, QoS profile, and others.

The customer thinks in terms of the service specifications. The customer's service specifications are formulated as the requirements and design

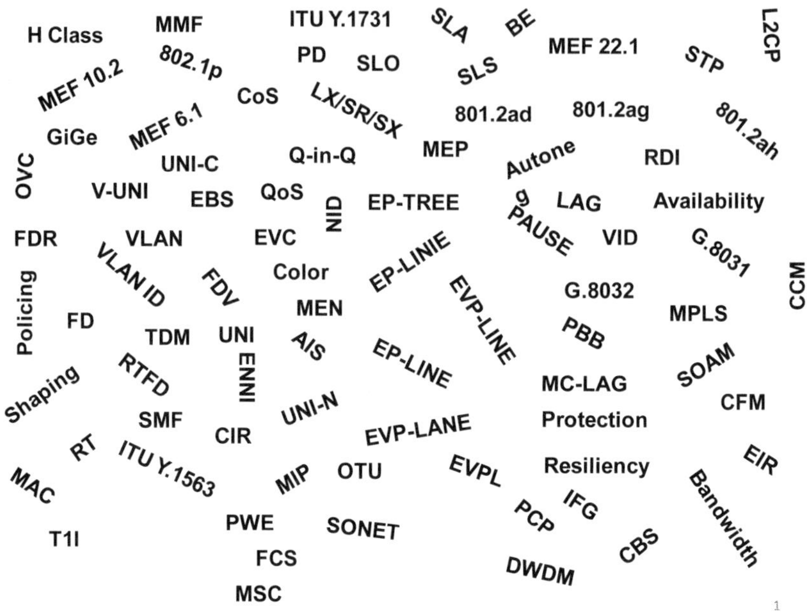

Figure 2.1 Ethernet concept soup.

assumptions in a more or less free format and structure. Whatever the format and structure of these requirements are, the complete specification must address the following design areas:

- The service logic defining service architecture;
- Transport technology defining the networking context for the service;
- Reliability and protection of the service;
- Quality of service;
- Service performance;
- Service management and verification;
- Physical aspects of interconnectivity between the customer's and provider's networks (quite often called engineering requirements).

The customer's and provider's views must meet; they do after all describe the same service (so we hope). The customer must be able to translate his service concepts into the Ethernet technology concepts, understand the impact and import of different elements of technology on his service, and develop the design specifications or requirements. The design specifications must be clear enough so that the Ethernet service provider will be able to understand what is required of her and construct MEBH service using service

Table 2.1
EVC and UNI Attributes for MEBH Service

Object Attributes	Service Attribute	EVC	Cell Site UNI	MTSO UNI
UNI attributes	Physical medium	NA	—	—
	Speed		—	—
	Mode		—	—
	MAC layer		—	—
	UNI MTU size		—	—
	Service multiplexing		—	—
	Bundling		—	—
	All to one bundling		—	—
	CE-VLAN ID for untagged and priority tagged service frames		—	—
	Maximum number of EVCs per UNI		—	—
	Ingress bandwidth profile per UNI		—	—
	Egress bandwidth profile per UNI		—	—
	Layer 2 control protocols processing		—	—
EVC attributes	EVC type	—		
	UNI List	—		
	Maximum number of UNIs	—		
	EVC MTU	—		
	CE-VLAN ID preservation	—		
	BMU frame delivery	—		
	L2CP	—		
	EVC performance	—		
	CE-VLAN COS preservation	—		
	CE-VLAN ID/EVC map	NA	—	—
	Ingress bandwidth profile per EVC		—	—
	Ingress bandwidth profile per CoS ID		—	—
	Egress bandwidth profile per EVC		—	—
	Egress bandwidth profile per CoS ID		—	—

objects and attributes meeting the customer's MEBH service specifications. The service provider must be able to relate the service objects and their attributes to the service specifications and understand from her side which attributes are critical to the service and how they have to be defined to accommodate the service requirements.

The provider's view of the service (based on attribute sets of the service objects) and the customer's view of the services (based on the functional service requirements) are in a way like two different reflections of the same underlying fabric over which the service is constructed. This dependency of views and reality is somewhat reflected in Figure 2.2.

We will discuss in more detail both views, but let us begin with service elements or concepts common to perspectives—Ethernet frame, UNIs, and EVCs.

2.2 Common Concepts

There is a set of common concepts fundamental to any perspective on the Ethernet services. These are the concept of the Ethernet frame, UNI, and EVC. We need to understand them before we can discuss any further design aspects.

2.2.1 Ethernet Frame

Ethernet frame is a basis of any Ethernet service. Understating the Ethernet frame format will facilitate the understanding of the Ethernet-based services as a lot of concepts related to the service are based on the specific fields in the

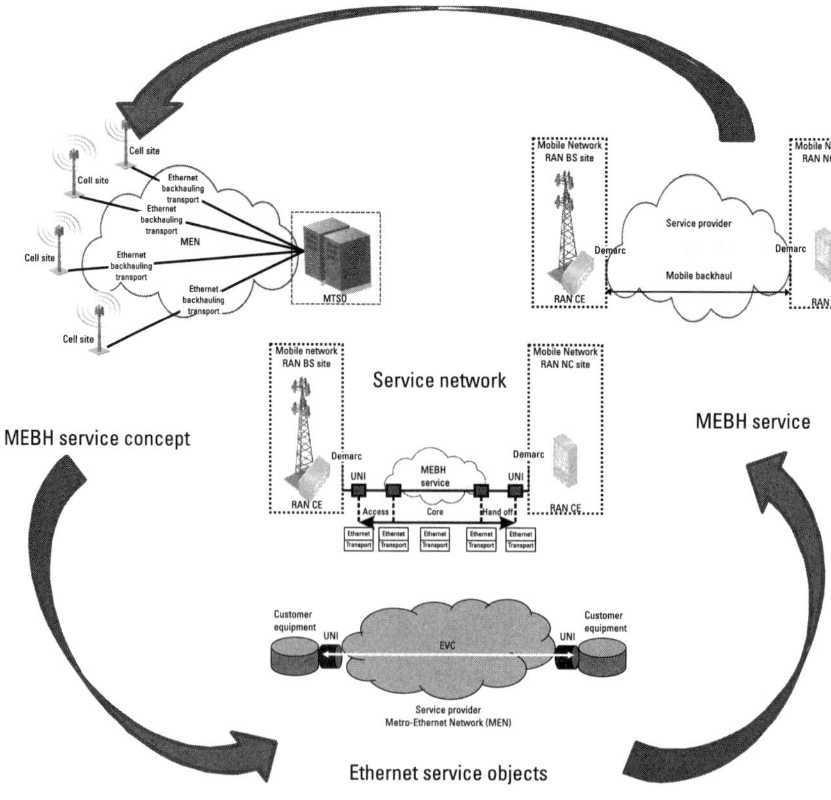

Figure 2.2 Service and network perspectives.

frame. Thus, knowing what the specific elements of the frame and its structure are will foster better understanding of the services characteristics.

Modern Ethernet services, including carrier Ethernet and MEBH, use the Ethernet frames as defined in the IEEE 802.3 standard. The Ethernet frame (specifically Ethernet II format) [3] composition is presented in Figure 2.3. The Ethernet II Frame in Figure 2.3 is called untagged, as there is no virtual LAN (VLAN) tag shim after the MAC SA field.

The Ethernet II frame has following fields:

- PRE: Preamble field of 7 octets [4];
- SFD: Start frame delimiter field of 1 octet;
- MAC DA: MAC destination address field of 6 octets;
- MAC SA: MAC source address field of 6 octets;
- EthType: EtherType field of 2 octets [5];
- Payload: Payload field of 46 to 1500 octets;
- FSC: Frame check sequence field of 4 octets (CRC);
- IFG: Interframe gap field of 12 octets.

The PRE, SFD, and IFG fields, all together adding 20 bytes, are considered the Ethernet protocol overhead. The minimum size of the Ethernet frame is 64 bytes. The maximum size is 1518 bytes. The payload is between 46 and 1500 bytes.

The tagged Ethernet frame has a VLAN tag shim of 4 octets added after the MAC SA field. The VLAN tagged frame is presented in Figure 2.4.

The VLAN tag field depicted in Figure 2.5 is composed of TPID, PCP, CF, and VID fields. The last three fields are grouped into the TCI field:

- TPID: VLAN tag protocol identifier of 2 octets. It is set to 0x8100 to identify the tagged frame.

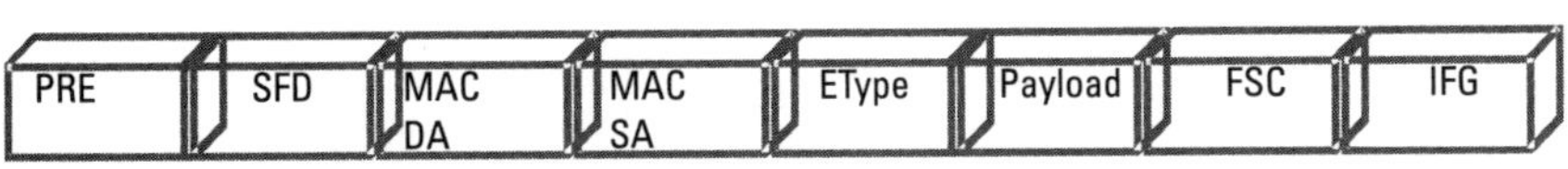

Figure 2.3 Ethernet II frame format.

Figure 2.4 Ethernet tagged frame.

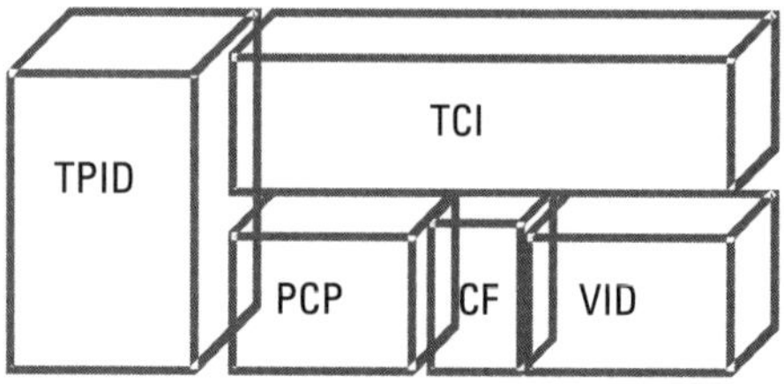

Figure 2.5 VLAN tag.

- TCI: VLAN tag control identifier containing three fields.
- PCP: Priority code point (PCP) of 3 bits containing the class of service (COS) designation of 0 to 7.
- CFI: Canonical form indicator of 1 bit; it can indicate drop priority (DE) [6].
- VID: VLAN identifier (VID) of 12 bites, with values from 0x000 to 0xFFF (value 1 to 4094 can be used for service; values 0x000 and 0xFFF are reserved). The VID = 0 denotes the frame that does not belong to any VLAN and is called a priority tag.

The Ethernet frame with two VLAN tags, the so called double-tagged frame, has two VLAN tags as represented in Figure 2.6. The outer tag is called S-tag (service provider TAG) and the TPID of this tag is 0x88a8, but this marking is not always applied to double tagged frames. The inner tag is called C-tag (a customer VLAN tag).

The format of a tagged Ethernet frame is defined in the IEEE standard 802.1Q. The format of the double-tagged Ethernet frame is defined in the IEEE standard 802.1ad [7].

More than two VLAN tags may be stacked together, but these constructs are not part of a standard.

One must differentiate between two concepts: Ethernet frame and service frame. The Ethernet frame is a frame with all fields as specified in Figures 2.3 through 2.6. The service frame, the MEF term, is a part of the Ethernet frame excluding PRE, SDE, and IFG fields. This differentiation is quite significant when talking about the Ethernet service and Ethernet service properties, and in particular about throughput, bandwidth, and bandwidth parameters.

Figure 2.6 Ethernet Q-in-Q frame.

The Ethernet frame has a maximum payload size of 1500 bytes. However, many applications require larger payload sizes. For these applications, so called jumbo frames have been created with the payload size of 9000 bytes, and even up to 9600 bytes. The 9000-byte limit is somewhat arbitrary; it came from the limited accuracy of the CRC field—a CRC loses effectiveness above 12,000 bytes and from the size of 8 KB of some applications datagrams (data storage and bulk data transfer). (The effectiveness of the CRC algorithm is expressed in the number of bit inversions in the network message that can be detected; the CRC algorithm used in the Ethernet frames for messages longer than 91607 bites cannot detect multiple bit inversion.) Quite often, jumbo frames are referred to as 9K frames, but this may mean either a payload size of the frames or the size of the full frames. In practice, a 9K frame may refer to 9000, 9216, or some other number of octets. A note of caution—the use of jumbo frames in the service may impact the performance of traffic on lower bandwidth interfaces, burst limits, and possibly latency and latency variations [8]. Thus, the offering of the jumbo frame service needs to be carefully thought through.

2.2.2 UNI and EVC

One may begin the job of explaining the constructs of the service in many ways. But it seems logical to start with the very basic, the fundamental—UNI and EVC—as presented in Figure 2.7.

The *user network interface* (UNI) is a demarcation point between the customer and service provider networks. The UNI is not a physical point in the network. It designates the part of the network where the customer and the provider of services meet. The UNI must be dedicated to one customer (i.e., it is not a facility shared by multiple customers) [9].

The *Ethernet virtual connection* (EVC) is an association between two or more UNIs [10]. The EVC establishes a virtual path along which frames that enter the EVC at one UNI—the ingress UNI—would travel to the other UNI or UNIs—the egress UNI. One UNI may have multiple EVCs associated with

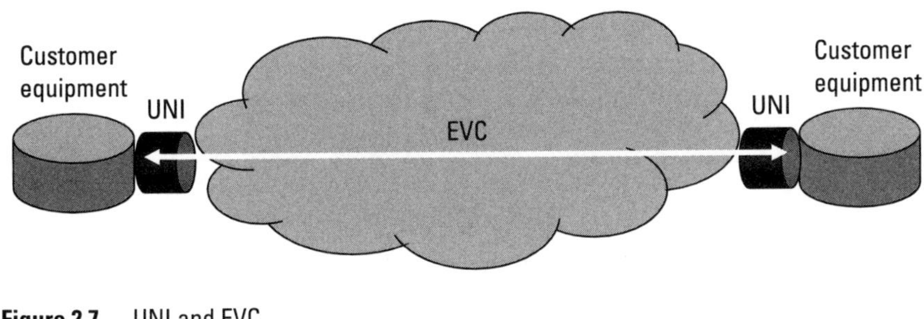

Figure 2.7 UNI and EVC.

it. This property of the UNI is called multiplexing. The multiplexing is achieved by "virtualizing" EVCs. In practical terms, virtualization means that an EVC is associated with a unique logical flow that can be distinguished from other flows and treated differently. The flow in multiplexed interfaces is primarily identified by the construct called VLAN ID, but may also be defined by other fields in the Ethernet frame [11].

UNIs may or may not be multiplexed. The UNI that is not multiplexed is called the *all-to-one* bundled UNI. This means that all frames coming to the UNI and not rejected for some reason are allowed to pass through the UNI and are associated with a single logical flow. The example of the all-to-one UNI is provided in Figure 2.8. The important aspect of the all-to-one bundled UNI is that it is associated with one EVC only (i.e., one cannot associate multiple virtualized services (EVCs) with the all-to-one bundled UNI).

The multiplexed [12] UNI can have multiple EVCs associated with it. Each EVC is defined as a separate flow and may be associated different egress single UNI, or with multiple UNIs. Figure 2.9 illustrates the concept of the multiplex UNI.

All of the discussed constructs and concepts of EVC and UNI are defined in MEF 10.2.

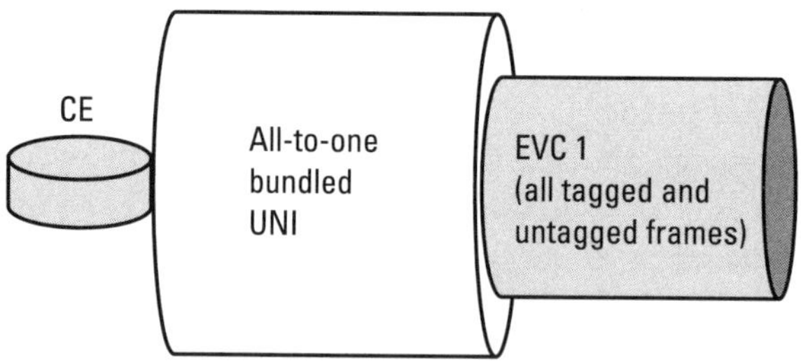

Figure 2.8 All-to-one bundled UNI.

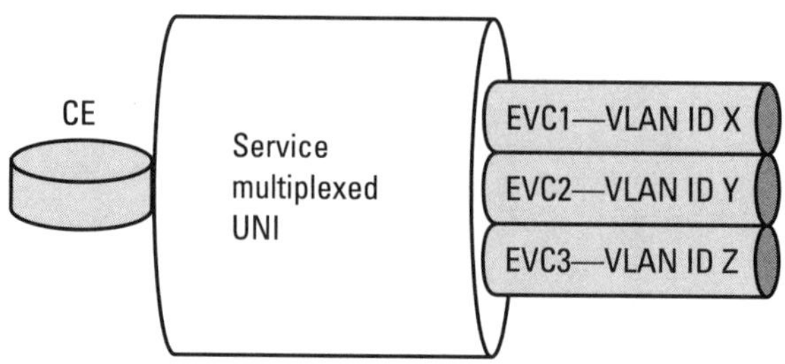

Figure 2.9 Multiplexed UNI.

2.2.3 ENNI and OVC

We have already briefly introduced MEBH services spanning multiple service providers as illustrated in Figure 2.10. Such services in addition to the EVC and UNIs have two or more OVCs and ENNIs—interfaces between OVCs.

Thus, the ENNI is an interface between two networks, connected with OVCs. An OVC is a logical connection between two or more UNIs and ENNI or between two or more ENNIs. The service frame that passed through the ENNI is referred to as the ENNI frame. The EVC in a case of multiple service providers is composed of multiple OVCs. The EVC in Figure 2.10 is composed of two OVCs (OVC-1 and OVC-2). Two UNIs face a customer CPEs, and the ENNI is between provider 1 and provider 2. The multiprovider EVCs with OVC and ENNI are defined in MEF 26.1.

With the basic constructs of the Ethernet service explained, we are ready now to look at both perspectives on the service—the customer's and provider's.

2.3 Provider's View

As we pointed out, the provider thinks about the service in terms of service constructs and its attributes. UNIs, EVCs, and their properties constitute for him the framework in which he defines the service. The provider begins with the specification of service functions and follows up with the service reference model.

2.3.1 Reference Models

Figure 2.11 presents the MEBH service architecture with overlaid basic carrier Ethernet service constructs. Thus, we have RAN BS and RAN NC locations interfacing through UNIs with the metro Ethernet network. The connection between UNIs, the logical path between them, is an EVC.

While the reference model represents a single MEN service, as discussed, it can be expanded to include multi-MEN services. Service specifications come together in a form of UNI, EVC, and QoS attributes sets. These are useful

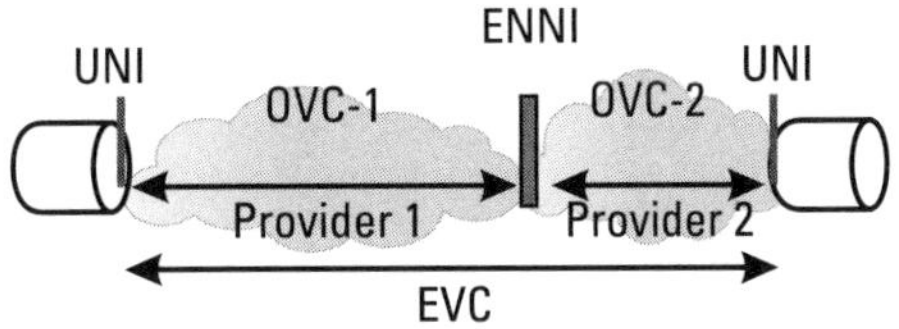

Figure 2.10 ENNI and OVC concepts.

Service Type	Port-Based (All to One Bundling)	VLAN-Based (EVC identified by VLAN ID)
E-Line (point-to-point EVC)	Ethernet Private Line (EPL)	Ethernet Virtual Private Line (EVPL)
E-LAN (multipoint-to-multipoint EVC)	Ethernet Private LAN (EP-LAN)	Ethernet Virtual Private LAN (EVP-LAN)
E-Tree (rooted multipoint EVC)	Ethernet Private Tree (EP-Tree)	Ethernet Virtual Private Tree (EVP-Tree)

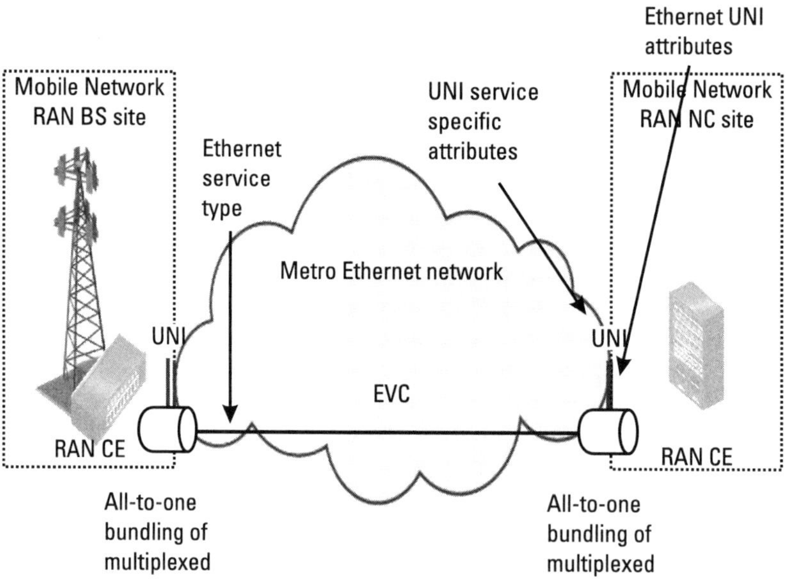

Figure 2.11 A high-level mobile backhauling architecture with basic CE constructs.

constructs, as they help to organize concepts into coherent, easy to refer to and analyze sets of objects.

2.3.2 UNI and EVC Attributes

The generic UNI and EVC attribute list is presented below. The attribute choice depends on the service supported on the UNI. Quite often the selection of the attributes, at least some of them, is a service orderable option. The list of attributes is taken after MEF 10.2. The following UNI and EVC attributes are defined:

- UNI identifier;
- Physical medium;
- Speed;
- Mode;
- MAC layer;

- UNI maximum transmission unit size;
- Service multiplexing;
- UNI EVC ID;
- CE-VLAN ID for untagged and priority tagged service frames;
- CE-VLAN ID/EVC map;
- Maximum number of EVCs;
- Bundling;
- All-to-one bundling;
- Ingress bandwidth profile per ingress UNI;
- Ingress bandwidth profile per EVC;
- Ingress bandwidth profile per class of service identifier;
- Egress bandwidth profile per egress UNI;
- Egress bandwidth profile per EVC;
- Egress bandwidth profile per class of service identifier;
- Layer 2 control protocols processing.

UNI identifier is usually an internal service provider parameter and should not be of concern to the wireless provider. The MEBH customer may have its own internal UNI ID. *Physical medium* describes the type of physical connectivity between the customer equipment and the service provider network and is subject to the collection of IEEE 802.3 standards. The speed attribute defines the physical speed of the UNI, and it may range from 10 Mbps to 10 Gbps (with 40 Gbps and 100 Gbps UNIs just around the corner as of December 2012). The *mode attribute* for CE services is full duplex. The *MAC layer* attribute for MEF services is defined by IEEE 802.3-2008 standard. The MTU size defines the size of the service frame that can be processed by the UNI. The MTU must at least accommodate the Ethernet frame of 1522 bytes. The *service multiplexing* attribute designates the type of the interface—all-to-one port or virtualized UNI (see preceding section for an explanation of UNI types). The *bundling* [13] attribute allows mapping of multiple VLANs to the single EVC. Such a mapping is usually defined by the C-VLAN to EVC map. Bundling means that flows with multiple VLAN IDs may be mapped to the same EVC. In effect, this is accomplished using the EVC VLAN maps. The *all-to-one bundling* attribute defines whether the UNI is virtualized (supporting multiple EVCs and services) or dedicated only to one EVC and service, so the value of this attribute must be in correlation with the UNI type. The *CE-VLAN ID for untagged and priority tagged service frames* attribute defines how untagged and priority tagged frame (VID=0) will be forwarded. The priority tagged farms are service frames with the VLAN ID =0. The *maximum number of EVCs* attribute specifies the number of EVCs on the UNI ranging from one to many for the multiplexed UNI. While the limit of EVCs per UNI is 4094, the provider will have his own rules that should be disclosed as part of the service definition. The *ingress and egress bandwidth profile per UNI* [14] *attributes* define parameters such as CIR, CBS, EIR, and EBS, and may include CM and CF parameters. CM is a flag for color-aware QoS mode. CF is a flag for coupling—a parameter used to control hierarchical policing in the MEF designed hierarchical policing schema. These parameters are defined in Section 3.4. The *layer 2 control protocols processing (L2CP)* rules are summarized in Section 3.1.5.

The EVC only attributes, defined in 10.2 as well, specify properties assigned to the EVC. These are as follows:

- EVC type;
- EVC ID;
- UNI list;
- Maximum number of UNIs;
- EVC maximum transmission unit size;
- CE-VLAN ID preservation;
- CE-VLAN CoS preservation;
- Unicast service frame delivery;
- Multicast service frame delivery;
- Broadcast service frame delivery;
- Layer 2 control protocols processing;
- EVC performance;
- EVC protection.

The *EVC type* attribute defines the type of the Ethernet service (or service logic) EVC is supporting—point-to-point, multipoint-to-multipoint, or rooted-multipoint EVC. The EVC ID attribute is an internal EVC designation, and it will vary from the provider to provider. The MEBH customer does not have to be aware of it. The UNI List is a list of UNI interfaces associated with the EVC. The list depends on the type of EVC. And so is the attribute *maximum number of UNIs* Attribute. For the point-to-point (p2p) EVC this limit is two, but for other EVC types this limit depends on the provider's policies (see Section 3.1, which defines the Ethernet service logic for the definition of Ethernet service types). The *EVC maximum transmission unit size* must of course be equal to or less than the UNI MTU size. The *CE-VLAN ID and CoS preservation* attributes define whether the customer VLAN tags and the PCP marking are passed transparently through the MEBH service. The attribute defining the pcp preservation is of the most critical importance to the customer and must be specified in the service description. The delivery of broadcast, multicast, and unknown unicast (BMU) frames should also be defined as a part of the service specifications. The L2CP per EVC policies may be different from per UNI policies, and they should be disclosed. The EVC *performance* attribute defines the performance parameters of the service over the EVC, including FTD, FDV, FLR, and Av. The EVC protection attribute is not the part of the MEF 10.2 specification. However, for MEBH services, the type of EVC protection, usually defined as availability per segment, per interface, or end to end, the failure time intervals, recovery time intervals, and protection topology is of critical importance and has to be defined.

Note: One has to have a proper view of EVC and UNI attribute sets defined by the MEF. The attributes themselves are features that are deemed to be the most important to the definition of the specific services or technologies. They come from standards defining physical and logical features of Ethernet services. It is not the case that the MEF defined them; they do exist by their own right as constituents of the technology. The role of the MEF has been to organize them into the collections and recognize their interdependencies. With MEF profiles, instead of searching for what needs to be defined, we just take a service object and its profile and have the attribute list ready. Thus, the role of the MEF profiles is that of formalization rather than definition. Why do we need to recognize it? It is because on some occasions the service needs to be designed in a way that violates some aspects of the MEF profiles. Such a service will work, if properly designed, but it will not be MEF compliant. Can this be done? Can non-MEF compliant services be implemented? Yes, of course, if some local service conditions may require such an approach. But such solutions should be avoided and discouraged [15].

2.3.3 ENNI and OVC Attributes

As with EVC and UNI, OVC and ENNI are also defined by a set of attributes. The complete specification of the ENNI and OVC should include:

- ENNI service attributes;
- OVC service attributes;
- OVC end point per ENNI;
- UNI service attributes;
- OVC per UNI service attributes.

We will limit this discussion to the ENNI and OVC attributes. The MEF 26.1 document provides the details of other elements of the ENNI. Table 2.2 provides the ENNI attributes. The OVC attributes are defined in Table 2.3. The presentation of the ENNI and OVC attributes is limited to tables only, as usually these are out of scope for the MEBH customer.

2.3.4 QoS Profiles

The QoS profile contains attributes that specify the quality of service. The quality of service, in general, may be a loose term that may include many diverse factors determining what we call a good service. The QoS profiles as defined here (after MEF) contain essential attributes that by the nature of the profile are measurable, quantifiable, and fundamental to the service specifications.

The QoS characteristics are collected in a QoS profile table, separate from EVC and UNI attribute tables, for the sake of convenience. As the QoS features are quite complex, it is easier to present them as a separate set and provide only references to them in the main EVC and UNI attribute set.

The example of the QoS profile in Table 2.4 has three generic QoS classes of service (CoS)—high (H), medium (M), and low (L). The QoS profiles with two classes, four or five could also be defined. The structure of the QoS profiles, regardless of the number of CoS classes, will stay the same. Each CoS class has four group of attributes:

- CoS class designation;
- Performance objectives for the class expressed in a metric of frame delay, packet loss, frame delay variation, and network (and service) availability.

Table 2.2
ENNI Attributes

Attribute Name	Summary Description
Operator ENNI identifier	An identifier for the ENNI intended for management purposes
Physical layer	The physical layer of the links supporting the ENNI
Frame format	The format of the PDUs at the ENNI
Number of links	The number of physical links in the ENNI
Protection mechanism	The method for protection, if any, against a failure
ENNI maximum transmission unit size	The maximum length ENNI frame in bytes allowed at the ENNI
End point map	The map that associates each S-tagged ENNI frame with an end point
Maximum number of OVCs	The maximum number of OVCs that the operator can support at the ENNI
Maximum number of OVC end points per OVC	The maximum number of OVC end points that the operator can support at the ENNI for an OVC.

Table 2.3
OVC Attributes

Attribute Name	Summary Description
OVC identifier	An identifier for the OVC intended for management purposes
OVC type	An indication of the number of OVC end points associated by the OVC
OVC end point list	A list of OVC end points
Maximum number of UNI OVC end points	The bound on the number of OVC end points at different UNIs that can be associated by the OVC
Maximum number ENNI OVC end points	The bound on the number of OVC end points at ENNIs that can be associated by the OVC
OVC maximum transmission unit size	The maximum length in bytes allowed in a frame mapped to an OVC end point that is associated by the OVC
CE-VLAN ID preservation	A relationship between the format and certain field values of the frame at one external interface and the format and certain field values of the corresponding frame at another external interface that allows the CE-VLAN ID value of the UNI service frame to be derived from the ENNI frame and vice versa
CE-VLAN CoS preservation	A relationship between the format and certain field values of the frame at one external interface and the format and certain field values of the corresponding frame at another external interface that allows the CE-VLAN CoS value of the UNI service frame to be derived from the ENNI frame and vice versa
S-VLAN ID preservation	A relationship between the S-VLAN ID value of a frame at one ENNI and the S-VLAN ID value of the corresponding frame at another ENNI
S-VLAN CoS preservation	A relationship between the S-VLAN PCP value of a frame at one ENNI and the S-VLAN PCP value of the corresponding frame at another ENNI
Color forwarding	The relationship between the color of an egress ENNI frame and the color of the corresponding ingress ENNI frame or service frame
Service level specification	Frame delivery performance definitions and objectives for frames between external interfaces
Unicast service frame delivery	Describes how ingress frames mapped to an OVC end point with a unicast destination MAC address are delivered to the other external interfaces with OVC end points associated by the OVC
Multicast service frame delivery	Describes how ingress frames mapped to an OVC end point with a multicast destination MAC address are delivered to the other external interfaces with OVC end points associated by the OVC
Broadcast service frame delivery	Describes how ingress frames mapped to an OVC end point with the broadcast destination MAC address are delivered to the other external interfaces with OVC end points associated by the OVC

Note: The metrics used in Table 2.4 are ITU defined metrics.

- Bandwidth profile (defined by specifying CIR and EIR).
- PCP values for traffic class, and depending on the number of QoS classes there are, the allocation of PCP code points may change. One may

find in other publications different marking schemas. However, the CoS markings in Table 2.4 conform to the general industry practices.

Table 2.4 should be read as follows: each QoS class designed by the specific CoS label (H, M, or L) has defined a set of class-specific performance parameters (H, M, L class) for four metrics (FTD, FDV, FLR, Av), a set of bandwidth parameters (CIR, CBS, EIR, EBS), and a coding of the PCP bits in the VLAN header. The designation of a range of the PCP values (2–3 or 0–1) allows the design of the specific class of traffic with any of the PCP values from the range or the use of the values to code the colored traffic.

2.4 Customer's View

The provider thinks about the service in terms of service functions. He perceives the service more as a unified, single object: a construct that does certain things and has certain properties. This view is akin to describing a car but not worrying about the details of the internal design or coming to a restaurant and ordering the meal but not caring about the kitchen stuff, at least, in the conceptualization phase of the meal. As with the meals that you eventually need to eat, so is the case with the networking or MEBH service; at one point you need to use the service. And then the details of the kitchen work (and the quality of the cook) come to the fore with force.

2.4.1 Reference Model

The customer's view of the service is expressed in terms of functional requirements or service specifications, as illustrated in Figure 2.12. Despite the apparent lack of organization, Figure 2.12 presents the way services are quite often conceptualized and represented.

Table 2.4
Generic QoS Parameters in Ethernet Services [16]

CoS Label	Frame Transfer Delay (FTD)	Frame Delay Variation (FDV)	Frame Loss Ratio (FLR)	Availability (AV)	Bandwidth Profile Parameters EIR=0	PCP CoS-Coding
H	H_{FTD}	H_{FDV}	H_{FLR}	H_{AV}	CIR>0,CBS, EIR=0, EBS	5
M	M_{FTD}	M_{FDV}	M_{FLR}	M_{AV}	CIR>0, CBS, EIR≥0, EBS	2-3
L	L_{FTD}	L_{FDV}	L_{FLR}	L_{AV}	CIR≥0, CBS, EIR≥0, EBS	0-1

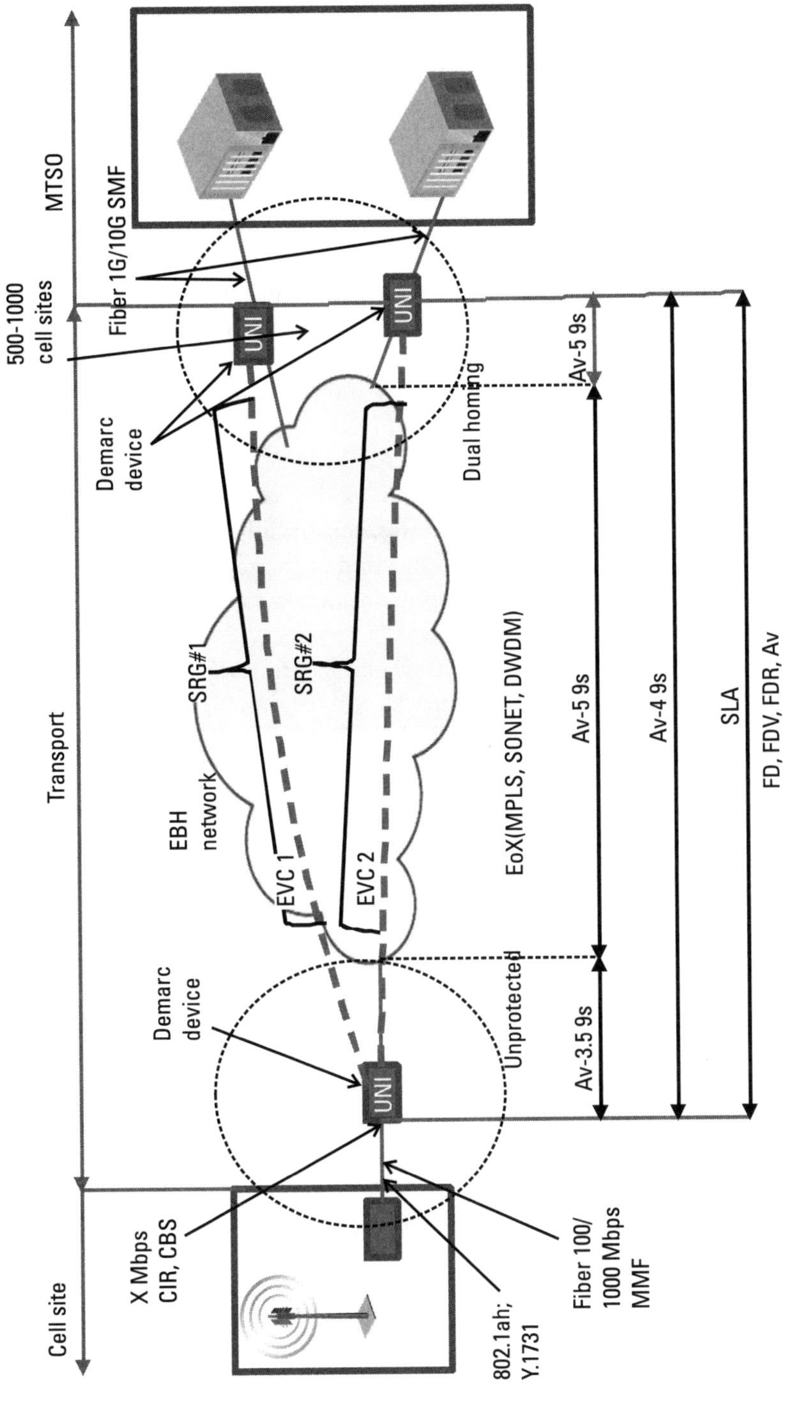

Figure 2.12 Customer's view of MEBH.

The elements of the diagram are intentionally not explained at this point to illustrate certain confusion of terms and ideas that the design process must unravel.

Around this type of diagram, using it as a reference, the service requirements for the service are developed. In other words, and this may come as a surprise, this type of diagram is the way the customer often conveys his requirements and conceptualize the service.

The customer sees his service in terms of functions or service features. There is no one agreed-to format for such requirements (that is why the MEF proposed service objects with attributes defining the most common service specifications). The functional groups proposed earlier—service logic; transport characteristics; reliability and protection; performance; quality of service, service management, and physical connectivity and verification—provide containers for most of the requirements needed to define the Ethernet (and MEBH) service. Other classifications of service requirements into different functional groups may be possible and are frequently used. None is better or worse. As long as they are logical and comprehensive, they fulfill their objectives, which is the complete and unambiguous specification of the service features. Arguing for one classification over another (as some readers would probably do) is a bit of a pedantic (in a negative way) exercise.

Thus, the headings of the functional groups proposed here may be different from one design to another. Yet, they all will (or should) contain the similar or even the same requirements. Let's look at each of the functional groups in more detail.

2.4.2 Service Functional Groups

Service logic defines logical connectivity between cell site UNIs and the MTSO UNIs. Quite often service logic is addressed under the title of service architecture or service topology. The service logic defines the type of Ethernet virtual circuits, the type of UNIs, the flow differentiation methods, and implicitly the format of the Ethernet service frame, filtering of control protocols, and a number of differentiated flows. All of these specifications are fundamental to the service, and all together they define the service architecture. All other requirements ride over the design specified by the service logic. If one gets the service logical design wrong, his service will not function properly.

Service transport technology provides the specifications of the networking context that the Ethernet service is provided. This includes the consideration of the protocol stacks in access and core of the transport network, as well as the underlying technologies. Despite the network designers mantra that the service should be seen as a "black box" by the customer, the customer should be aware of the technology used to deliver the service, as in some cases technology itself will force certain solutions and limit available design options.

Service protection and resiliency defines how a service responds to failures of its components. While the customer should be satisfied with stating the service requirements, the selection of certain solutions to the service architecture is dictated by the desired resiliency of the service. Understanding what protection is and how it is obtained and controlled is an essential part of the design process.

The *quality of service* features discussed under this heading include the definition of QoS classes and their attributes such as traffic profiles and performance targets. This set of requirements includes also the specification of mechanisms used to enforce QoS.

Service performance includes the definition of the measurable properties of the service. These are expressed as metrics such as latency, packet loss, jitter, and availability. There is of course more to these specifications that just naming metrics and their targets. Many metrics have similar definitions and many ways they may be measured and interpreted. Any ambiguity in this functional area will make the service performance intractable and unverifiable.

Service management and verification includes the features of service that inform about the status of the service and signal service malfunctions. Included are the service, operation, administration and maintenance functions (SOAM) and the service level agreement (SLA).

Service interconnectivity (often referred to as engineering requirements) includes the physical aspects of the interconnectivity between the service and the customer networks at the meet point (cell site or MTSO) such as (but not limited to) physical requirements of interfaces, connectivity media, fiber, copper, dimensioning of interconnectivity equipment, power, and environmental specifications.

2.4.3 MEBH Service Requirements

We can now put together a full set of service requirements in each of the functional groups for the MEBH service (customer view). These requirements are presented in Table 2.5.

2.5 Common Grounds

Despite the differences between these two approaches to the specification of services (the service provider's approach based on the attributes sets of service objects and customer's approach based on functional groups), the specifications have to meet at one point; they obviously have to as they define the same service! The objective of the design process is to connect these two views together

Table 2.5
Service Requirements

Functional Group	Requirements (MEBH Service)
Service logic	The following aspects of the service logic must be defined: EVC type—connectivity between UNI at the cell site and UNI at the MTSO; Number of EVCs on UNI cell site; Number of EVC at MTSO; BMU processing; Reserved VLAN IDs; Processing of untagged frames; VLAN IDs; L2CP at UNIs.
Transport	The following aspects of the transport must be defined: Access technology requirements; Transport technology requirements; MTU.
Protection	The following aspects of protection and resiliency of the service must be defined: Target availability for EVC, UNI, and MTSO UNI; Convergence times for different failure modes; Shared risk node or link groups; Transport protection.
Quality of service	The following QoS aspects of the service must be defined: Bandwidth required; Bandwidth profiles; Signaling of class of service; Policing and/or shaping requirements.
Service performance	The following aspects of service performance must be defined: Metrics for each class of service; Metric targets.
Service management and verification	The following aspects of the service verification must be defined: SOAM architecture; SLA measurements methodology; Metrics reporting methodology; Fault notification.
Interconnection engineering specifications	The following aspects the engineering specifications of the service must be defined: Types of permissible interfaces at UNIs; Required fiber types at UNIs; Space, environment requirements; Permissible types of power feed at the interconnection location.

and create the single vision of the service. One may say that the MEBH service provider and the customer speak different languages about the same problems. For the customer to get the service she wants, she needs to understand what the provider is telling her through the service attributes, and for the service provider to provide the desired service, he needs to understand what really the customer means by her requirements.

The customer needs to understand what all service attributes mean, not in the abstract, but with regard to her service. She needs to project her functional requirements onto the Ethernet service objects and their attributes. The mapping of service attributes into the service functional groups or service specifications is somewhat reflected in Table 2.6.

Table 2.6 may point to the source of the communication problem between the customer and the provider—the attributes assigned to EVC and UNI do not map one to one into the service requirements functions used by the customer. Objects and attributes for the EVC and UNI may be allocated across the service requirements groups in few ways, and every such mapping—even the tentative mapping in this table—is open to interpretation. Yet, the common element between these two sets of requirements is the the service and the underlying network; a prolonged argument about what attributes go where is missing the main point—how to construct the best architecture for the service. In the presence of reality, extended discussions on the formal aspects of classifications of features and requirements describing this reality seem like a bit of waste of time if it is not in the standards forum.

One word of caution—the reader should not take from this section the impression that the problem of communicating the service requirements to the service providers, if it happens, are related to the fact that the service provider tends to see the service as a series of attributes and the customer as a set of functional requirements. Such an impression would be a gross oversimplification of the issue at hand. If such problems happen, there is a multitude of factors that play in, not just an attributes-requirements dichotomy. However, the juxtaposition of attributes versus functional requirements is meant here as a representation of two points of view that two sides (customer and service provider) involved in the design tend to have due to the different roles they take in the design process, and while there is a grain of truth in it, it should not be taken 100 percent literally.

There is no one, best way to forge the smooth communication between the customer and the provider short of actually going the design process together. This book is such an attempt. The proposed approach is to walk through an

Table 2.6
Service features and Attributes

	EVC Attributes	UNI Attributes	QoS profiles	SLA Features
Service logic	X	X	—	—
Transport characteristics	X	X	—	—
Reliability and protection	X	X	X	X
Quality of service	X	X	X	X
Performance	—	X	X	X
Service verification	—	—	X	X
Interconnectivity	—	X	—	—

example of the service and look at both sides of the design and see how they fit together.

To connect both approaches to the design specifications (attributes' view and the functional grouping of service requirements), we will develop a complete example of the MEBH architecture for a hypothetical use case of the MEBH service. The architecture will not reflect any specific MEBH operator. Yet, it will contain all the critical components that should comprise such a service and may serve as an example of how to develop the requirements and specifications for real-life design cases.

In the sections that follow we will attempt to navigate through the maze of the Ethernet service concepts to arrive at the working MEBH service architecture; this voyage is somewhat intended to address the issue signaled in Figure 2.1. In the journey we use the conceptual structure of service functions used by the customer to organize the Ethernet service specifications (as we said, any way to organize the service concepts is good as long as it is logical and complete). With this framework as our reference we will explain the generic Ethernet service concepts, their dependencies, and their specific uses in the MEBH service. In the final sections of the book we use discussed concepts to constructs the design for the MEBH exemplar service.

Endnotes

[1] Robert Lloyd, the engineer at the IBM Advanced Computing Systems Division, talking about the microprocessor, c. 1968. Quoted in Cerf, C., and N. Navansky, *The Experts Speak*, New York: Pantheon Books, 1984.

[2] There is no shortage of books that tell you what, but there is a limited number of books that would tell you why and how. Good "what" introduction to Ethernet technology can be found in Halabi, S., *Metro Ethernet*, Indianapolis, IN: Cisco Press, 2003, and more up to date and complete in Kasim, A., *Delivering Carrier Ethernet: Extending Ethernet Beyond the LAN*, New York: McGraw-Hill, 2008.

[3] Ethernet II is called sometimes Ethernet 802.3.

[4] One octet is 8 bits or 1 byte

[5] Ethertype field is used to indicate which protocol is encapsulated in the payload of an Ethernet frame.

[6] This field—canonical format indicator (CFI)—was initially used to designate a MAC address in canonical format (MAC address in standard notation or canonical format is written with transmission bit order with the least significant bit transmitted first) and was always set to zero. The Ethernet port receiving a frame with a CFI bit set to one would not bridge the frame to the untagged port or drop the frame. The CFI bit was used for compatibility between Ethernet and token ring (IEEE 802.5) networks.

[7] The list of 802.1 IEEE standards is provided in the Appendix.

[8] The concepts of burst and performance are explained in Section 3.4.1.

[9] MEF 10.2, Section 5.

[10] Ibid.

[11] An Ethernet frame flow is a set of frames associated with a given connection or connectionless stream having the same source MAC address (SRC), the same destination MAC address (DST) in case of point-to-point connections or the same set of destinations in case of point-to-multipoint connections, the same Ethernet VLAN tag, or the same session identification (e.g., IP addresses or port numbers from a higher-layer protocol). The VLAN tag includes both a VLAN ID and the user priority bits, commonly referred to as the p-bits (auth. p-bits are encoded in the field priority code point, or PCP, in the Ethernet tagged header) [ITU Y.1563.]

[12] "Multiplexing" in essence means a combination of many higher order entities, logical flows in this case, into the lower order aggregates, UNIs in this case. One can combine IP flows with the port numbers into one IP flow. One can also combine lower-order flows into the higher order aggregates.

[13] "When a UNI has the bundling attribute, it *must* be configurable so that more than one CE-VLAN ID can map to a particular EVC at the UNI. The bundling service attribute is independent of the EVCs at the UNI. An EVC with more than one CE-VLAN ID mapping to it *must* have the CE-VLAN ID preservation service attribute and the list of CE-VLAN IDs mapped to the EVC *must* be the same at each UNI in the EVC. Bundling is compatible with service multiplexing." MEF 10.2 Sec 7.9.

[14] "The bandwidth profile defines the set of traffic parameters applicable to a sequence of service frames. Associated with the bandwidth profile is an algorithm to determine service frame compliance with the specified parameters. In the case of an ingress bandwidth profile rate enforcement is accomplished via the disposition of noncompliant service frames." MEF 10.2.

[15] "The solution described in a technical standard is seldom the only solution to the problem it addresses. In fact, it often is not even the best solution to the problem (complex engineering problems tend not to have a 'best solution,' but only a least inconvenient tradeoff). All standards do identify a single point in the solution space of the problem, with the goal of establishing a benchmark for interchangeability and interoperability." Beck, M., *Ethernet in the First Mile*, New York: McGraw Hill, 2005.

[16] MEF 23.1.

3

Service Functions Technology Overview

Worthless.
—Sir Airy, 1842, regarding the analytical engine [1].

The following sections discuss basic Ethernet concepts and the Ethernet service elements in each of the service function groups:

- Service logic defining connectivity, flow definition, and interface types;
- Service transport technology;
- Service reliability and protection;
- Service quality of service;
- Service performance;
- Service management and verification;
- Interconnectivity between the customer and providers networks.

3.1 Service Logic

Service logic defines the way UNIs are interconnected (i.e., the logic of the service, the flow of the traffic, and the connectivity between edge of the services or UNIs). More importantly, each type of Ethernet service has certain properties that define how the service operates. Thus, each service has distinguishing characteristics and is targeted for the specific tasks (read service). Table 3.1 summarizes the types of Ethernet services.

Table 3.1
MEF Ethernet Services [2]

Service Type [3]	Port-Based Service	VLAN-Based (Virtualized Service)
E-line (point-to-point [p2p] EVC)	Ethernet private line (EPL)	Ethernet virtual private line (EVPL)
E-LAN (multipoint-to-multipoint [mp2mp] EVC)	Ethernet private LAN (EP-LAN)	Ethernet virtual private LAN (EVP-LAN)
E-tree (rooted multipoint [p2mp] EVC)	Ethernet private tree (EP-TREE)	Ethernet virtual private tree (EVP-tree)

3.1.1 E-Line Service

The E-line service is a service in which each EVC has associated exactly one pair of UNIs. The E-line service maybe port-based (EP-line) or virtualized (EVP-line). In the Ethernet private line (EPL) service, each UNI has associated only one EVC. Thus, each UNI is associated only with one other UNI. All service frames [4] that ingress at one UNI should be transported to the other UNI. The E-line EPL service architecture is depicted in Figure 3.1 and the related EVC UNI map in Table 3.2

In the EVPL E-line service, each UNI may have associated multiple EVCs and multiple UNIs. However, one EVC is still associated with a pair of UNIs. Each EVC is a considered a separate flow and may have different characteristics defined in its EVC profile. Subsequent sections will detail these aspects of the EVCs. All frames ingressing at one UNI into one EVC will egress at the other UNI associated with this EVC. Frames that are not associated with a specific

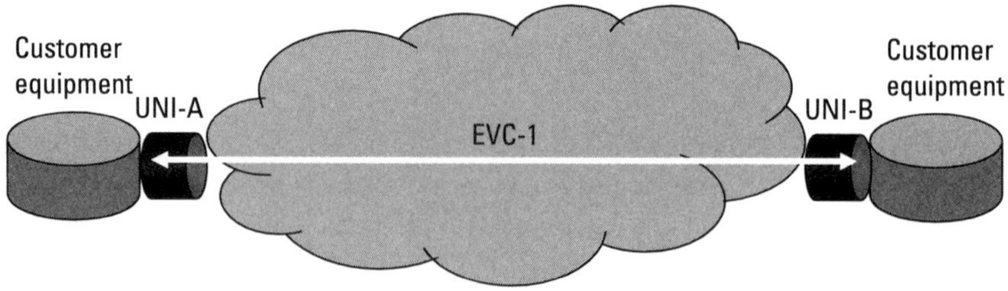

Figure 3.1 EP-line port-based service.

Table 3.2
EVC UNI Map for the EP-Line Service

EVC	UNIs
EVC 1	UNI A UNI B

EVC should not be admitted to this EVC. This condition achieves a virtual [5] separation of the flows in each EVC; hence the name—virtual connection. This separation property resembles the dedicated physical circuit of TDM technology. However, in the case of EVC the circuit, the connection is virtualized (i.e., logical, not physical). The EVPL service architecture is depicted in Figure 3.2.

The EVC-UNI map [6] for the EVPL service contains also (as EPL) one EVC two UNIs as in Table 3.3.

3.1.2 E-LAN Service

The E-LAN services associate one EVC with multiple UNIs. The EP-LAN services can be port-based (EP-LAN) or virtualized (EVP-LAN). E-LAN services are depicted in Figure 3.3 and Figure 3.4.

The EP-LAN service has different operational properties from EVP-LAN services that need to be understood by the customer buying this service. In the EP-LAN services frames entering one of the UNIs will be transferred to another UNI on this EVC based on the frames MAC addresses. The network supporting such service must allow frames to be switched to the specific destination egress UNI based on the frame's MAC DA. In the Ethernet technology the function allowing this switching is called bridging. How and where the bridging function is implemented in the network is dependent on the underlying

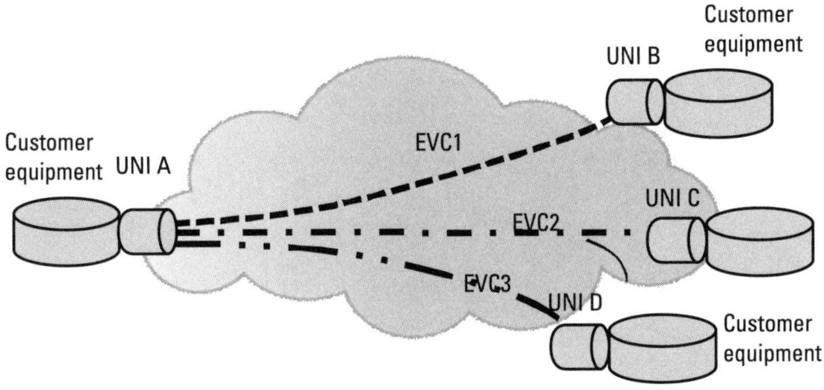

Figure 3.2 EVPL service.

Table 3.3
EVC Map for the EVPL Service

EVC	UNIs
EVC1	UNI A UNI B
EVC2	UNI A UNI C
EVC3	UNI A UNI D

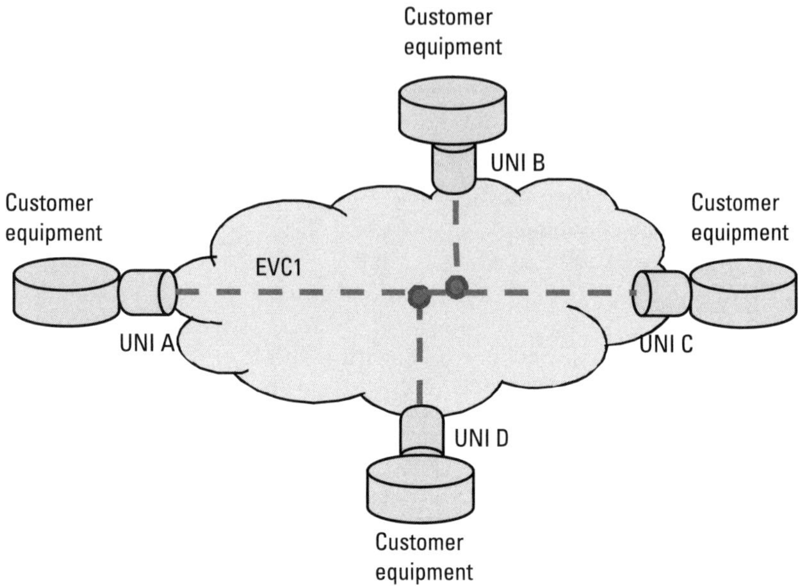

Figure 3.3 EP-LAN service.

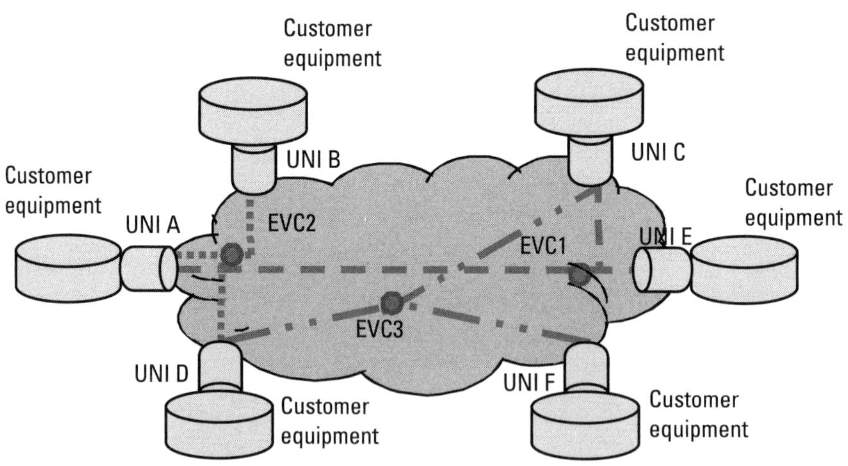

Figure 3.4 EVP-LAN services.

Ethernet technology, and it should be of lesser concert to the customer buying MEBH service.

The EP-LAN service (Figure 3.3) is a port-based service, which means that all frames (tagged or untagged) arriving at a UNI may be transported to their destinations, if permitted to enter EVC. There is only one EVC associated with the UNIs. On Figure 3.3 the EP-LAN EVC connects four UNIs. In this service, service frames can travel from any UNI to any UNI in the EVC. Thus, broadcast, multicast, and unknown unicast (BMU) frames are sent to all UNIs. Known unicast frames will be sent to their specific destinations, once these are learned. The EP-LAN service, as with all bridged services, requires careful engineering to prevent BMU [7] storms. The EVC UNI map for the service in Figure 3.3 is illustrated in Table 3.4.

The E-LAN service offered over a multiplexed UNI, denoted as the EVP-LAN service, is depicted in Figure 3.4.

The EVP-LAN service is a virtualized EP-LAN service (i.e., the service allows the coexistence of different EP-LAN-EVCs on the same UNI). As presented in Figure 3.4 on UNI A and UNI D, there are two EVP-LANs. Each EVP-LAN has a unique EVC associated with it. Most of the properties of the EP-LAN service are preserved in the EVP-LAN service. Service frames coming to the UNI from the customer side will be mapped to the specific EVP-LAN EVC based on the EVC map. Frames not present in the EVC map will be dropped. Untagged frames, if no mapping specifies the EVC on which they could be transported, will be dropped. The EVP-LAN service is a port-multiplexed service; more than one EVC may exist on one port, and one EVP-LAN EVC UNI map is illustrated in Table 3.5.

The EVP-LAN can be combined with any other virtualized service as depicted in Figure 3.5. In Figure 3.5 the EVP-LAN service coexists with EVPL on UNI A. As in the EP-LAN service, in the EVP-LAN service UNIs belonging to the specific EVP-LAN can send frames (BMU) to any of the UNIs participating in the EVP-LAN. On the specific UNI only frames with the VLAN IDs that are associated with the EVP-LANs associated with this UNI (via EVC maps) will be transported. All other frames as well as untagged frames will be dropped, unless special provisions are made to map such frames to the specific EVP-LAN EVC.

Table 3.4

EP-LAN EVC UNI Map

EVC	UNIs
EVC1	UNI A, UNI B, UNI C, UNI D

Table 3.5

EVP-LAN EVC UNI Map

EVC	UNIs
EVC1	UNI A, UNI C, UNI E
EVC2	UNI A, UNI B, UNI D
EVC3	UNI D, UNI C, UNI F

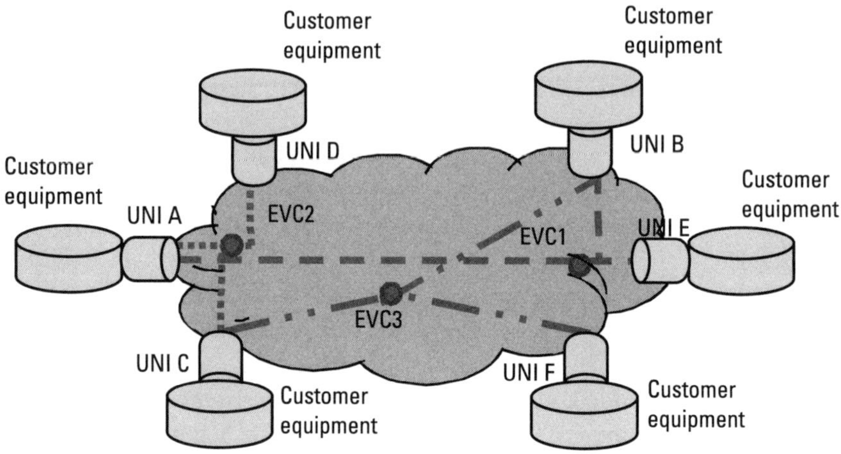

Figure 3.5 EPV-LAN coexisting with EVPL service on the same UNIs.

In combining the diffident virtualized service, like EVP-LAN and EVPL, on a single UNI the properties of the EVCs are preserved. This means that the traffic in each EVC is separated, each EVC is either p2p or mp2mp, and each EVC may have different attributes. The EVC UNI maps define which frames are mapped to which EVC and which are dropped. Each EVC is also a separate broadcast domain containing the BMU traffic. The EVC UNI map for the EVP-LAN and EVPL service is presented in Table 3.6.

3.1.3 E-Tree Service

The E-tree is the third generic Ethernet service supported in MEF Ethernet networks. As with other services, the E-tree service maybe port-based (EP-TREE) or virtualized (EVP-TREE). The EP-tree service is represented in Figure 3.6.

In the EP-tree service one UNI is defined as a root of the tree, and other UNIs are defined as leaves. The root UNI can send any traffic to any of the leaves. The leaf UNIs can send traffic only to the root UNI. Thus, leaves cannot communicate between themselves. In such a service the BMU traffic is greatly reduced, as the traffic of such type between leaves UNIs is blocked. The root UNI has therefore a critical role in the EP-tree service, as it is only through

Table 3.6
EVP-LAN and EVPL Service EVC UNI Map

EVC	UNIs
EVC1	UNI A, UNI C
EVC2	UNI D, UNI F
EVC3	UNI A, UNI B, UNI E

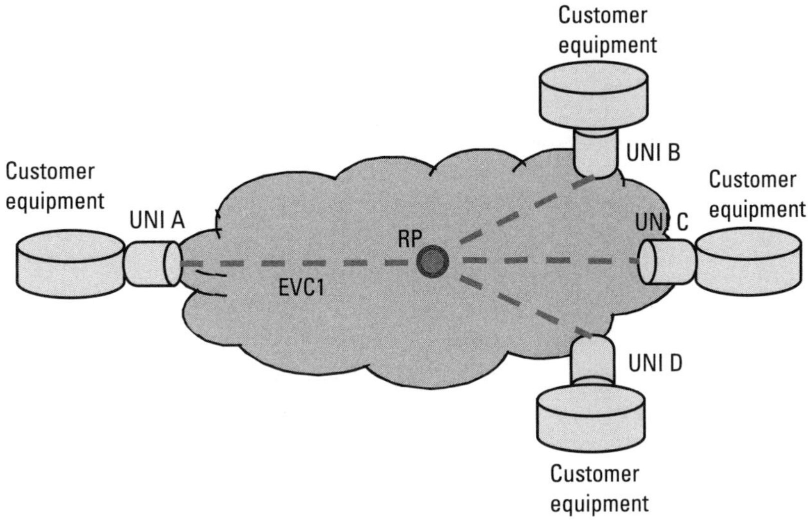

Figure 3.6 EP-tree service.

this UNI that leaf UNIs can communicate. The EP-tree EVC UNI map is presented in Table 3.7

To increase the resiliency of the service, one may implement the EP-tree with two or more roots (as in Figure 3.7). In such a construct, called multiroot tree, if the primary root UNI fails the other root UNI can take over the functions of the primary root, preserving the continuity of service.

As with all Ethernet services, the EP-TREE service can be virtualized as the EVP-tree. The EVP-tree construct allows coexistence of multiple virtualized Ethernet services on the same UNI as illustrated in Figure 3.8; one could design multiplexed EVP-trees or a combination of other virtualized services. Table 3.8 presents the EVC UNI map for this service.

The EVP-tree services preserve the E-tree properties and behavior (with the exception of the all-to-one bundling) and can be, as with other multiplexed services, combined with other virtualized Ethernet services. An example of such a combination of virtualized services is presented in Figure 3.9. The service EVP-LAN as EVC-2 is implemented on the same UNIs as the EVP-tree EVC-1. Table 3.9 provides the EVC UNI map for the service. Other combinations of virtualized services are also possible.

Table 3.7
EP-Tree EVC UNI Map

EVC	**UNIs**
EVC 1	UNI A (root), UNI B (leaf), UNI C (leaf), UNI D (leaf)

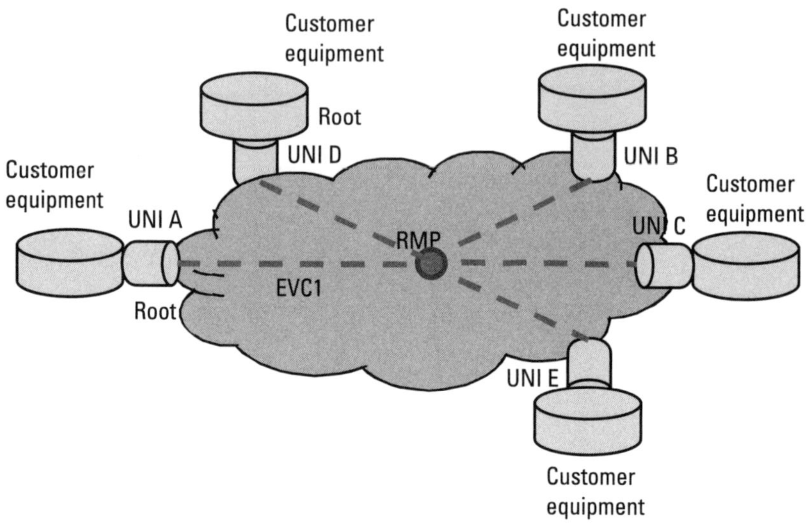

Figure 3.7 Multiroot EP-tree.

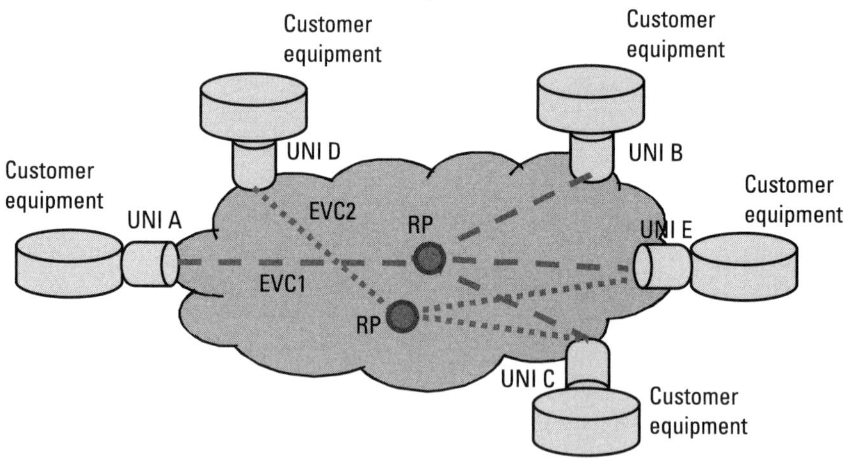

Figure 3.8 EVP-tree service.

Table 3.8
EVP-Tree EVC UNI Map

EVC	UNIs
EVC 1	UN A (root), UNI B (leaf), UNI C (leaf), UNI E (leaf)
EVC-2	UNI D (root), UNI C (leaf), UNI E (leaf)

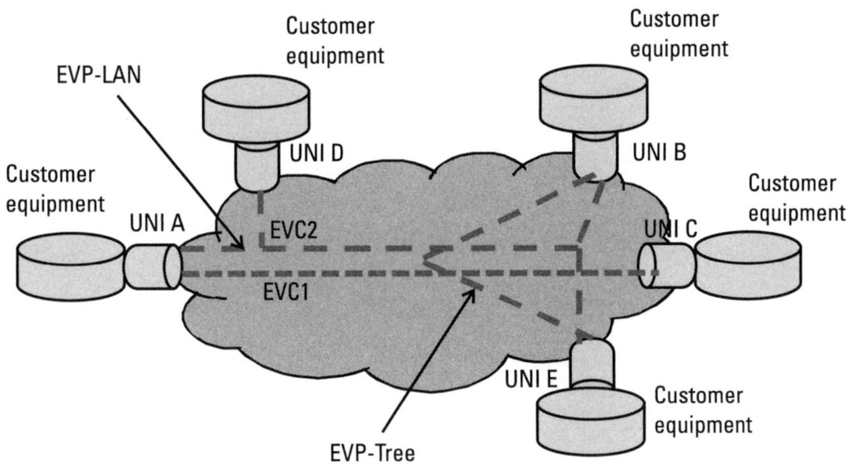

Figure 3.9 EVP-LAN and EVP-tree coexisting on the same network.

Table 3.9
Mixed EVC EVC UNI Map

EVC	UNIs
EVC1	UNI A (root), UNI B (leaf), UNI C (leaf), UNI E (leaf)
EVC2	UNI A, UNI D, UNI B, UNI E

3.1.4 Services in Multiprovider Architecture

The canonical Ethernet service types (E-LAN, E-tree, and E-line) are also available in the multiservice provider architecture. The definitions of the services do not change. What changes is the detailed design of the service, depending on the location of specific OVCs and UNIs. Next, we present examples of two canonical services, E-line and E-LAN, from the previous section in the multiservice provider environment. Figure 3.10 presents the EVP-line service over the ENNI, and Figure 3.11 presents the EVP-LAN service over ENNI. The presented services are deployed over two service providers. But they can be in a similar way extended over multiple providers. The design of the multiprovider service is, or should be, transparent to the customer of the MEBH service. The definitions of the ENNI related service attributes is a responsibility of the providers of the OVCs.

The service in Figure 3.10 uses three EVPL EVCs. The EVCs span two service providers; each EVC is therefore composed of two OVCs. The mode by which the service providers interface with each other should be transparent to the customer. The providers would interface with a single ENNI rather than with three, one per EVC, as depicted in Figure 3.10. Each EVC is p2p. They

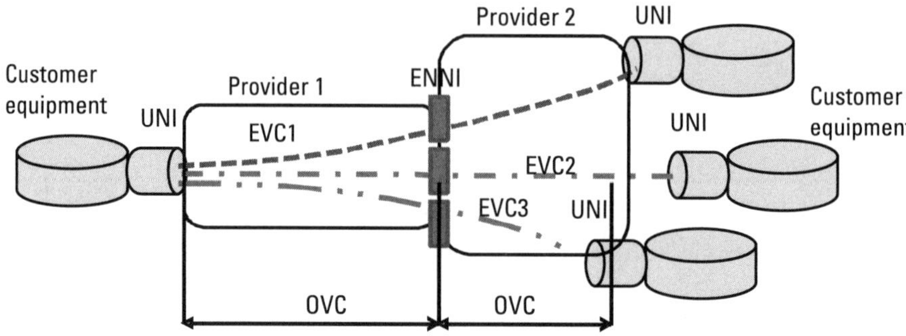

Figure 3.10 EVP-line service over ENNI.

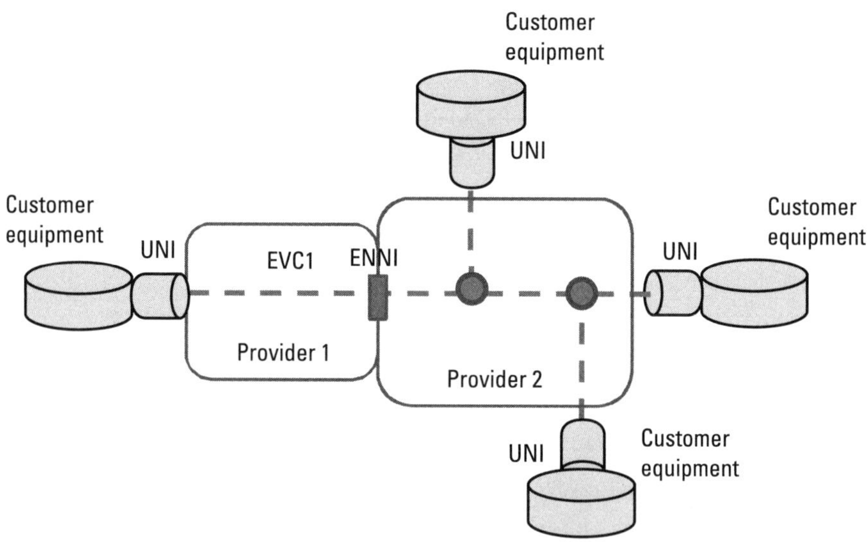

Figure 3.11 EP-LAN service over ENNI.

share on the near side one UNI, but each one has a different UNI on the far side. Each EVC may have different attributes.

Figure 3.11 represent the EP-LAN service over the ENNI. The service is offered by two providers. The providers are interfacing with a single ENNI.

The service characteristics of the specific service type should not change, regardless of whether the service is provided over the single or multiple providers [8].

3.1.5 Layer 2 Control Protocol (L2CP)

L2CP filtering rules define the characteristics of the EVC and UNI, and, as they are the function of the service logic, they are discussed in this section.

The Layer 2 Control Protocol (L2CP) term refers to the traffic flows that carry the information about the status of services or network [9]. The L2CP frames have a MAC destination address (DA) within the range 01-80-C2-00-00-00 through 01-80-C2-00-00-0F and 01-80-C2-00-00-20 through 01-80-C2-00-00-2F. The treatment of L2CP frames is addressed by standard IEEE 802.1ad-2005 Provider Bridge. Three actions are defined for the L2CP frame on the UNI: *tunnel*, *peer*, or *discard*.

Discard means that the UNI will discard ingress L2CP frames [10]. Peer means that the MEN will actively participate with the protocol [11]. For example, LACP/LAMP, link OAM, port authentication, and E-LMI might be peered by the UNI. Tunnel means that service frames containing the protocol will be transported across the MEN to the destination UNI(s) without change [12]. What follows is a simplified description of the L2CP processing rules as stated in MEF 6.1.1 [13].

The MEF 6.1.1 is mostly concerned with the L2CP frames falling within the 01-80-C2-00-00-00 to -0F MAC DA. The control protocols with their Ethertype using these MAC DAs are listed in Table 3.10.

These L2CP protocols are standard protocols defined by the standard forums with well-specified properties and with applications to the Ethernet services in general, not for the specific platform. One word of caution: sometimes the boundary between the standard protocol and the vendor-specified protocol may be a bit fuzzy. Some vendors created protocols that later on became standardized (like E-LMI).

The action (tunnel, peer, or discard) for each L2CP service frame will be decided using a two-step logic based, first, on the frame's MAC DA and then on the frame's Ethertype and subtype or LLC code. The logic for processing of the L2CP service frames is presented in the flow chart in Figure 3.12.

Thus, if for the specific frame, based on the frame's MAC DA, Table 3.11 mandates tunneling, the frame must be tunneled. If for this frame, based on

Table 3.10
L2CP Control Protocols in MEF 6.1.1

Protocol Type	Ethertype/Subtype
STP/RSTP [14]/MSTP [15]	NA [16]
PAUSE [17]	0x8808
LACP|LAMP	0x8809/01|02
Link OAM	0x8809/03
Port authentication [18]	0x888E
E-LMI [19]	0x88EE
LLDP [20]	0x88CC
PTP Peer-Delay [21]	0x88F7
ESMC [22]	0x8809/0A

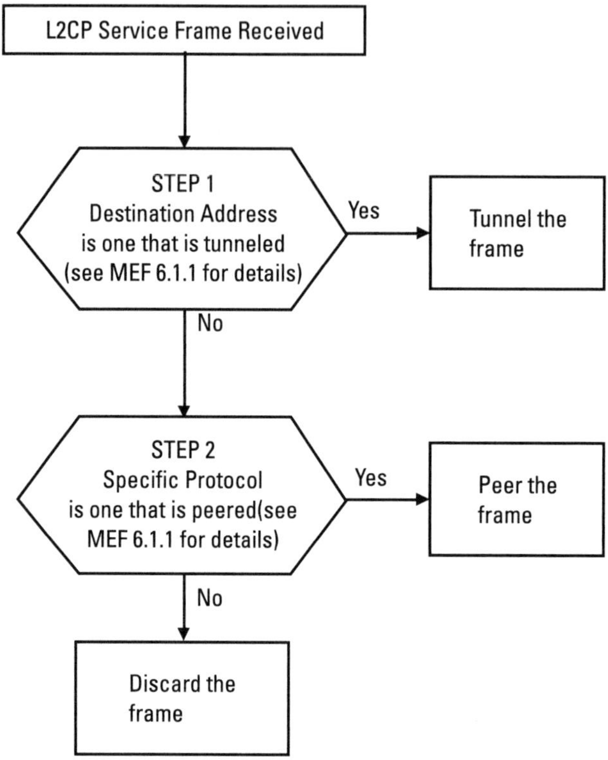

Figure 3.12 L2CP processing steps [23].

Table 3.11
L2CP Processing—Step 1

Destination MAC Address	L2CP Action for EPL, EP-Tree, EP-LAN	L2CP Action for EVPL, EVP-Tree, EVP-LAN
01-80-C2-00-00-00 [24]	*Must* tunnel	*Must not* tunnel (additional requirements may apply as per the specific service type)
01-80-C2-00-00-01 through 01-80-C2-00-00-0A	*Must not* tunnel (additional requirements may apply as per the specific service type)	
01-80-C2-00-00-0B	*Must* tunnel	
01-80-C2-00-00-0C	*Must* tunnel	
01-80-C2-00-00-0D	*Must* tunnel	
01-80-C2-00-00-0E	*Must not* tunnel (additional requirements may apply as per the specific service type)	
01-80-C2-00-00-0F	*Must* tunnel	

the frame's MAC DA, Figure 3.12 mandates peer or discard, the action for the L2CP frame is based on the frame's protocol type (defined by the Ethertype and subtype or LLC code) and is specified in subsequent, service specific tables—Table 3.12 and Table 3.13.

Outside of this group of protocols there is a whole gamut of protocols whose treatment is not defined so precisely. These protocols fall into the category of vendor-specific protocols, and their processing on the UNI will differ from platform to platform. For the detailed account of the L2CP processing as specified by MEF, the reader should consult the standard.

The processing of the L2CP traffic on the ENNI is the subject of the ongoing work in MEF (as of 2013).

3.2 Service Transport

3.2.1 Ethernet Technology

Ethernet technology is specified by the IEEE standards 802.3 and 802.1. The IEEE 802.3 [25] defines the technology that supports the IEEE 802.1 [26] architecture. In rare cases, a network engineer designing the Ethernet service deals with the Ethernet technology only; usually the Ethernet service is offered in the complex networking environment. Thus, to fully understand the environment in which the Ethernet service is delivered, it is necessary to provide even a limited view of the networking context of the Ethernet services. The selection of the technology over which Ethernet service is delivered will have a definite

Table 3.12

L2CP Processing Requirements for Virtualized Ethernet Services

Protocol Type	L2CP Action EVPL	L2CP action EVP-LAN	L2CP action EVP-Tree
STP/RSTP/MSTP	*Must* peer on all UNIs or discard on all UNIs	*Must* peer on all UNIs or discard on all UNIs	*Must* peer on all UNIs or discard on all UNIs
PAUSE	*Must* discard on all UNIs	*Must* discard on all UNIs	*Must* discard on all UNIs
LACP/LAMP	*Must* peer or discard per UNI	*Must* peer or discard per UNI	*Must* peer or discard per UNI
Link OAM	*Must* peer or discard per UNI	*Must* peer or discard per UNI	*Must* peer or discard per UNI
Port authentication	*Must* peer or discard per UNI	*Must* peer or discard per UNI	*Must* peer or discard per UNI
E-LMI	*Must* peer or discard per UNI	*Must* peer or discard per UNI	*Must* peer or discard per UNI
LLDP	*Must* discard on all UNIs	*Must* Discard on all UNIs	*Must* discard on all UNIs
PTP peer delay	*Must* peer on all UNIs or discard on all UNIs	*Must* peer on all UNIs or discard on all UNIs	*Must* peer on all UNIs or discard on all UNIs
ESMC	*Must* peer or discard per UNI	*Must* peer or discard per UNI	*Must* peer or discard per UNI

Table 3.13
L2CP Processing Requirements for Port-Based Ethernet Services

Protocol Type	L2CP Action EPL	L2CP Action EP-LAN	L2CP Action EP-Tree
STP/RSTP/MSTP	*Must* peer on all UNIs or discard on all UNIs	*Must* peer on all UNIs or discard on all UNIs	*Must* peer on all UNIs or discard on all UNIs
PAUSE	MUST discard on all UNIs	*Must* discard on all UNIs	*Must* discard on all UNIs
LACP/LAMP	*Must* peer or discard per UNI	*Must* peer or discard per UNI	*Must* peer or discard per UNI
Link OAM	*Must* peer or discard per UNI	*Must* peer or discard per UNI	*Must* peer or discard per UNI
Port authentication	*Must* peer or discard per UNI	*Must* peer or discard per UNI	*Must* peer or discard per UNI
E-LMI	*Must* peer or discard per UNI	*Must* peer or discard per UNI	*Must* peer or discard per UNI
LLDP	*Must* peer or discard per UNI	*Must* discard on all UNIs	*Must* discard on all UNIs
PTP peer delay	*Must* peer or discard per UNI	*Must* peer on all UNIs or discard on all UNIs	*Must* peer on all UNIs or discard on all UNIs
ESMC	*Must* peer or discard per UNI	*Must* peer or discard per UNI	*Must* peer or discard per UNI

impact on the services—it may limit its service features or allow for its smooth growth and evolution.

The context of networking technologies is usually presented in a form of protocol stack, or functional layers as seen in Figure 3.13 [27]. The protocol stack is an abstract representation of networking protocols and their physical realizations.

Each layer in the stack has a set of defined functions and interfaces, or service access points (SAPs), to the layers above and below. Layers communicate only through these interfaces by passing up or down the stack PDUs. The

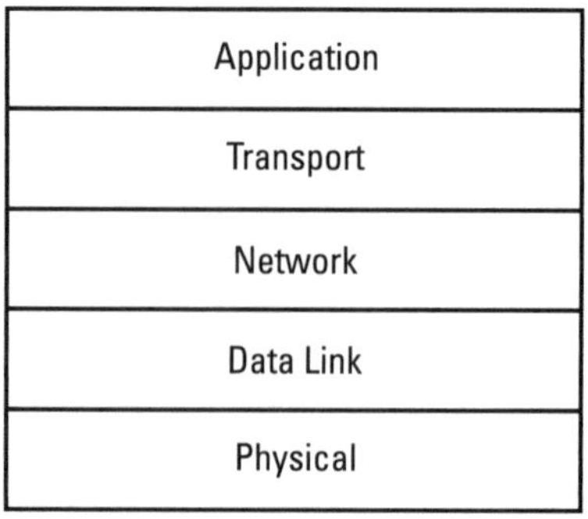

Figure 3.13 Simple protocol stack.

isolation of layers allows implementation of the functions on the specific layer independent of other layers.

We can briefly characterize each of the layers as having the following functions. Application layer functions are responsible for user applications and interfaces. The transport layer is responsible for flow control among other functions. The network layer is responsible for packet forwarding and routing. The data link layer is responsible for packet forwarding between network entities. The physical layer is responsible for physical (hardware) encoding and decoding as well as transmission management.

The IEEE Ethernet standards define physical and logical aspects of Ethernet technologies in the data link and physical layers defined as in Figure 3.14. The Ethernet service uses facilities and functions of the data link layer—layer 2. That is why Ethernet service is called a layer 2 service.

For the purpose of this book it is sufficient to provide only the brief characteristics of each of the IEEE 802.3 layers. The physical layer defines how the bits are transfer over the physical (wireline or wireless) media. It also specifies the physical connectivity or interfaces on the devices connected to the Ethernet. The data link layer from the generic protocol stack is composed of the media access control (MAC) and logical link control (LLC) sublayers. The MAC sublayer supports data encapsulation, media access management and transmission control, and other functions. The LLC layer primarily supports flow control and multiplexing or frames, and it provides the interface to upper protocol layers [28].

Native Ethernet uses the IEEE Ethernet protocol stack—physical and data link layers. In carrier networks, however, Ethernet service may be present in several locations in the stack depending on the design of the service and transport. In addition, the Ethernet service context may be different in the

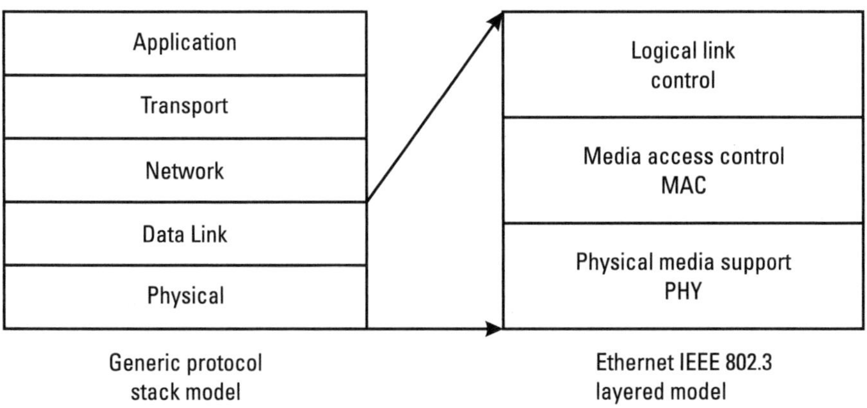

Figure 3.14 Ethernet IEEE 802.3 layered model.

access, handoff, and core segments of the MEBH service as depicted in Figure 3.15.

Figure 3.16 presents the network design in which the Ethernet service is delivered over the MPLS [29] transport network over the Ethernet transport. The service provider will interact only with the lower Ethernet layer, in most cases. The provider has to be aware of the design (or a composition) of the transported frame, as quite often unexpected problems in the service may have a root cause in the lower layer design. Of course, the MPLS protocol stack in Figure 3.16 is greatly simplified. The detailed protocol stack would list all of the MPLS components and complexities of transport.

The service architect must be cognizant of the Ethernet frame context for many reasons. One is that the networking technology in which Ethernet services are embedded impacts the way the services behave—among aspects impacted are performance, resiliency, fault propagation, restoration, and QoS. Why is that so? It is because each technology has its own control plane, management plane, and signaling plane methods and resources. And these specificities must be recognized to fully understand the end-to-end service. The following sections introduce the most common networking technologies in which the Ethernet service is delivered.

3.2.2 Transport Technologies

The networking technologies employed with Ethernet service are presented in Figure 3.17. Ethernet may be delivered over wireline or wireless media. We leave out the wireless technology from further discussion and focus on wire-

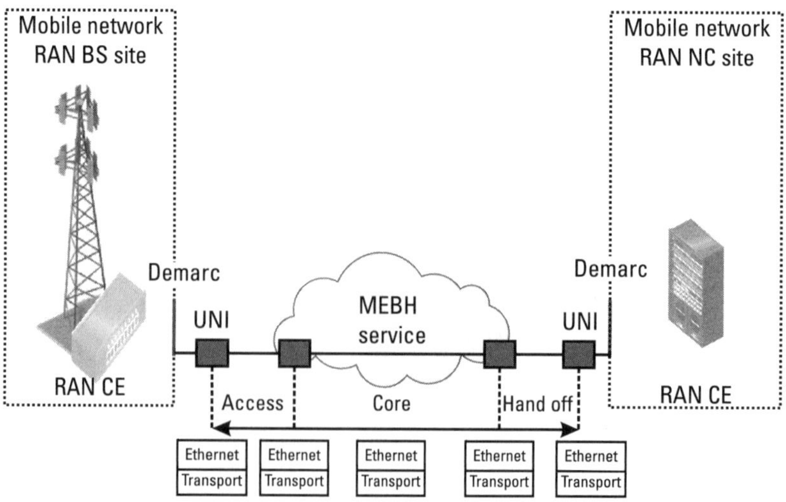

Figure 3.15 Architecture for delivering Ethernet services.

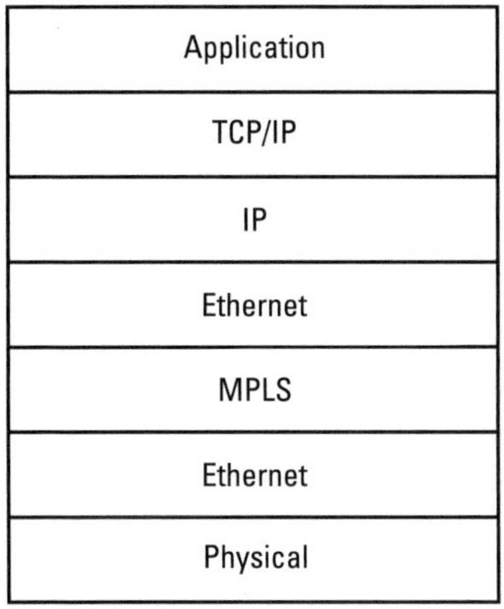

Figure 3.16 Ethernet Service over MPLS.

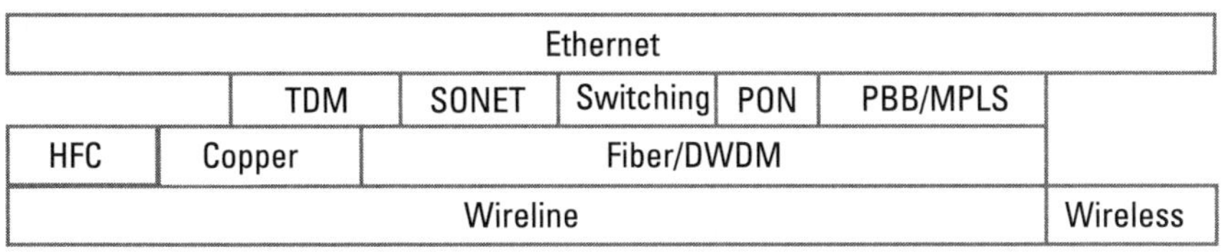

Figure 3.17 Ethernet delivery technologies.

line. Wired technology may support coaxial, copper, or fiber media. Over these media one may use TDM, SONET/SDH, switching, and PON technologies. Provider backbone bridging (PBB) and multiprotocol label switching (MPLS) provide layer 2 transport functions for the Ethernet technology mediating often between the lower transport layers and the Ethernet service. Thus, they are not equivalent to TDM, SONET/SDH, and PON.

The following descriptions of different technologies are far from being comprehensive and are intended only to highlight the most important features relevant, in author's view, to the MEBH service design, meaning the support for Ethernet service attributes. Such an approach, by its nature, leaves out vast amount of technical details and introduces simplifications in representing otherwise complex technologies—reader beware!

3.2.2.1 Ethernet over SONET/SDH (EoS)

EoS refers to the Ethernet service supported over the synchronous optical network (SONET/SDH)/ synchronous digital hierarchy (SDH) transport technologies. SONET is predominantly North American standard. SDH is used outside of North America. In the EOS architecture, Ethernet frames are sent over the SONET/SDH link encapsulated into the generic framing procedure [30] (GFP) block that maps the asynchronous Ethernet flows into the synchronous the SONET/SDH stream. GFP mapping is a generic mapping procedure that can be used to map packets into the SONET/SDH frames. The mapping drops the Ethernet frame control fields, improving the efficiency of the transport. The SONET/SDH technology provides the guaranteed bandwidth and robust protection mechanisms.

In SONET/SDH the main transport is implemented using the multiples of STS-1 containers of roughly 50 Mbps. The actual payload capacity is lower. For fined granularly (less than STS-1) one can use VT1.5 containers of 1.6 Mbps. The combinations of STS-1 or VT1.5 containers are called virtual concatenation groups (VCG) with STS1 combinations called high-order VCGs and with VT1.5 called low-order VCG. Examples of rates offered by EoS service are provided in Table 3.14.

The Ethernet bandwidth [32] or the actual Ethernet layer (layer 2) bandwidth depends on the packet size (service frame size). As the Ethernet over SONET/SDH uses GFP encapsulation (12 bytes), the actual payload bandwidth is reduced by the overhead percentage. For example, STS1 50.112 Mbps payload rate has to be reduced for 512 bytes frames by 512/(512+12) = 0.977 or 97.7 percent, giving 48.77 Mbps Ethernet or service frame throughput.

EoS guarantees high QoS quality of the service (no overprovisioning), robust protection architectures, robust OAM plane, and very fast (within 50 ms) recovery times. It is suited for EPL and EVPL (point-to-point) Ethernet services. At present (2013), the SONET/SDH technology is going out of favor. The main reasons for this are lack of bandwidth flexibility that is available with layer 2 technologies, relatively high cost of the infrastructure, no support for classes of service, lesser efficiency as compared to packet-based technologies, and no support for overprovisioning. All these limitations should not prevent anyone from seeing EoS technology service as delivering reliable and matured service.

Table 3.14
Examples of SONET/SDH Payload Rates

SONET/SDH Frame Format/ Optical Carrier Level [31]	SDH Level and Frame Format	Payload Bandwidth (Mbps)
STS-1/OC-1	STM-0	50.112
STS-3/OC-3	STM-1	150.336
STS-12/OC-12	STM-4	601.344
STS-48/OC-28	STM-16	2,405.376
STS-192/OC-192	STM-64	9,621.504
STS-768/OC-768	STM-256	38,486.016

EOS technology can be used both in the access and core transport segments of the MEBH service in a variety of topologies (e.g., point-to-point or ring). EoS technology is defined by ITU (SDH) and ANSI (SONET/SDH) standards [33].

3.2.2.2 Ethernet over Cable

EoCable or EthernetoHFC refers to the technology specified by data over cable service interface specification (DOCSIS) [34] for high-speed data transfer over the hybrid fiber-coaxial (HFC) media, which combines optical fiber and coaxial cable. Data signal in HFC technology is encoded over the radio frequency. The data signal is converted into the modulated RF signal and back by the modem device at the customer premises and in the head-end equipment on the other end, respectively. Cable industry typically uses 42–750-MHz RF range, 5–42 MHz for upstream data, and 54–860 MHz for downstream transmission as 6-MHz wide channels. A single 6-MHz channel can support multiple data stream or multiple users with layer 2 (LAN) protocols. Different modulation techniques are used, including quadrature phase shift keying (QPSK) upstream and quadrature amplitude modulation (QAM 64-256) downstream.

Management of different traffic flows is provided with QoS features introduced in DOCSIS 1.1. Depending on the DOCSIS release, the throughput (maximum usable throughput without the overhead) may range from 38 Mbps per channel or multiples of it ($n \times 38$ Mbps) in DOCSIS 3.0 downstream to 27 Mbps or multiples of it ($n \times 27$ Mbps). No maximum number of channels (n) is defined. The EoHFC network has a tree topology. Thus, the capacity of the connection is shared among the users. The amount of bandwidth available to the user depends on many factors, among them the number of users, type of traffic, and noise in the cable plant.

3.2.2.3 Ethernet over Wavelength Division Multiplexing (EoWDM)

Ethernet over WDM is a catch term that includes Ethernet transport over optical technologies such as EoOTU, EoDWDM, and Ethernet over fiber (EoF). Ethernet over optical transport network (OTU [35]) uses a new technology defined to optimize the transport of multiple service over the DWDM media. The OTU technology is specified in two ITU standards, ITU G.706 and G.798. It is referred to quite often as a digital wrapper, as it allows the transport of Ethernet, video, SONET/SDH, fiber channel, and others over the common transport unit (OTU) at different speeds ranging from 2.48 Mbps to 100 Mbps (OTU-1 at 2.7 Gbps, OTU-2 at 10.7 Gbps, OTU-3 at 43 Gbps, or OTU-4 at 112 Gbps). The OUT rates are provided in Table 3.15.

The OTU technology offers several advantages such as multiplexing of the client signals, improving the bandwidth efficiency, transparent encapsulation of a client signal (Ethernet traffic is encapsulated in to the GFP or GFP-T

Table 3.15
OTU Rates and Client Signals

OTU Type	OTU Rate (Gbps)	OTU Payload Rate (Gbps)	Client Signals
OTU1	2.6661	2.48832	STM-1/OC-3, STM-4/OC-12, STM-16/OC-48, GbE,
OTU1e	11.049	10.3215	10GbE LAN
OTU2	10.709	9.9953	STM-64/OC-192, 10GbE WAN, 10GbE LAN (GFP),
OTU2e	11.095	10.356	10GbE LAN
OTU3	43.018	40.150	STM-256/OC-768, 40GbE
OTU4	111.80997	104.35597	100GbE

frame), OAM facilities, and 50 msec restoration of transmitted signal. It is essentially point-to-point technology.

EoF refers to the Ethernet technology delivered over optical fiber in native format in the variety of interface media and fiber connector options—multimode and single mode fiber with a variety of speeds such as Fast Ethernet, 1GiGE, 10 GiGe, and higher.

EoDWDM refers to Ethernet over dense wavelength division multiplexing (DWDM) or packet-based transport technology over DWDM. EoDWDM uses OTU wrapping, offering more efficient use of the available bandwidth in comparison to TDM technology. DWDM technology extends from access to the core. By accommodating Ethernet in its native format and with combing of the layer 2 features, it allows for better grooming of traffic by mapping layer 2 flows directly into wavelengths. Additional advantages such as end-to-end management, monitoring, and reducing the complexity of equipment presents the EoDWDM as a less costly and more efficient alternative to the other transport solutions.

3.2.2.4 Ethernet over DSL

Digital subscriber loop (DSL) is a technology that adapts the existing copper-based connections for high-speed data access. The current DSL speeds are reaching past 100 Mbps. The limitation of the technology is its dependence on the distance. As the distance from the head-end office to the customer end point increases, the capacity diminishes significantly. Another feature of the DSL technology is its asymmetry, in particular in earlier released versions. High-end DSL speeds are supported over 10–14Kft, with the maximum speed supported over < 10 Kft distance from the CO. The DSL may reach up to 24 Kft but with significantly reduced bandwidth. DSL bandwidth dependency on the distance is heavily dependent on DSL technology. See Table 3.16 for details of DSL transport.

Table 3.16
Selected DSL Technologies [36]

Family	ITU	Name	Ratified	Maximum Speed
ADSL	G.992.1	G.dmt	1999	7 Mbps down; 800 Kbps up
ADSL2	G.992.3	G.dmt.bis	2002	8 Mbps down; 1 Mbps up
ADSL2plus	G.992.5	ADSL2plus	2003	24 Mbps down; 1 Mbps up
SHDSL (updated 2003)	G.991.2	G.SHDSL	2003	5.6 Mbps up/down
VDSL	G.993.1	Very-high-data-rate DSL	2004	55 Mbps down; 15 Mbps up
VDSL2 -12 MHz long reach	G.993.2	Very-high-data-rate DSL 2	2005	55 Mbps down; 30 Mbps up
VDSL2 -30 MHz short reach	G.993.2	Very-high-data-rate DSL 2	2005	100 Mbps up/down
Vectored VDSL2	G.993.5		2011	120 + Mbps

3.2.2.5 Ethernet over Passive Optical Networks (EoPON)

EoxPON technology refers to the class of access technologies called passive access technologies. The name comes from the use of the passive optical splitters in the network that enable the use of a single laser for several subscribers. The splitter distributes the signal among customer connections downstream (toward the customer) toward the optical network termination (ONT) unit at the customer premises. Upstream, the ONT uses the allocated time slots (TDM). The EoxPON technology has several variants including Ethernet-PON (EPON) [37], gigabit (GB) PON (GPON), 10-GB Ethernet PON (10GEPON) [38], or wave-division multiplex PON (WDM-PON). Prevailing installations are that of GPON technology [39]. Current GPON technology offer 2.5 Gbps toward the customer and 1.25 Gbps upstream. With a 32-fold splitter, this potentially may offer the up to 78 Mbps downstream and 39 Mbps upstream. The EPON technology may deliver the service over a 20-km range and with different splitters (16-fold or less), and the bandwidth to the customer may be increased even to 1 Gbps.

3.2.2.6 Ethernet over TDM

EoTDM refers to the Ethernet over TDM $n \times$ T1(DS1)/E1 (bonded T1), T3(DS3) (45 Mbps), or its derivatives. This technology is sometimes referred to as Ethernet over copper (EoC). The EoTDM is delivered over the twisted pair cable. A T1 circuit delivers 1.544 Mbps. With bonded technology (which essentially allows aggregating multiple T1 circuits, augmenting the available bandwidth in multiples of T1), one may bond up to eight T1s offering 12 Mbps. Above eight T1s, the bonding becomes less economic. Above 40–50 Mbps, it is more cost-efficient to move to fiber from copper-based technology. EoTDM is precisely the technology from which mobile providers are migrating. As a reminder, T1 and similar technologies haven proven to be outage prone and expensive, and thus not suitable for the demands and requirements of the MEBH service needed for LTE.

3.2.2.7 Ethernet over MPLS

Ethernet over MPLS [40] is not technology in the sense of SONET/SDH, OTN, TDM, or PON. It is a packet-based packet technology at the protocol stack at the 2.5 layer that offers to some extent client-agnostic, packet-based transport supporting aggregation, protection, and the rich set of SOAM functions. MPLS itself needs layer 2 (data link layer) and layer 1 (physical layer). Thus, it is often combined with the SONET/SDH, OTU, or Ethernet at layer 1 and 2. The EoMPLS architecture provides carrier-grade functions such as resiliency, protection, QoS, traffic engineering, and complex control plane and SOAM facilities not supported to the same extent by the Ethernet itself. EoMPLS was positioned as a competitive technology to the pure layer 2 tunneling architecture offered by provider backbone bridging (PBB) known as well as mac-in-mac architecture [41] EoMPLS and PBB designs could be considered in several, but not all, aspects functionally equivalent.

3.2.3 Comparison of Different Technologies

With the variety of networking technologies (SONET/SDH, OTU, HFC, xPON xDSL, fiber, TDM) that Ethernet can be delivered over in access and core of the network the customer has sometimes a difficult decision trying to understand the choices and their impact on the service.

Table 3.17 summarizes the discussed technologies and their most important attributes. There is a clear separation for technologies into those that can be used in access segment and those that can be used in the core and handoff. In access, Ethernet may be provided over xPON, xDSL, fiber, TDM, HFC, SONET/SDH, and of course EoF. Technologies in access differ significantly by the granularity and limitation of the bandwidth available. Most of the access technologies are point to point. Some of them support CoS classes, some provide only one class of service, and some would allow overprovisioning. In the core and the handoff segments, Ethernet may be provided over SONET/SDH, OTU, and fiber. These technologies scale up to over 10 GiGe, allowing different levels of aggregation and multiplexing. These technologies usually provide the protection (node and network) support < 100 ms restoration times. The transport technologies such as SONET/SDH and OTU may be enhanced by providing the packet awareness on the edges of the service or in the intermediate points. MPLS or PBB technologies add layer 2 features enhancing the service. However, they do require transport technologies underneath.

Figure 3.18 depicts the several modes of delivery of Ethernet services; the options are not exhaustive. As we indicated we have excluded wireless access,

Table 3.17
Ethernet Delivery Technologies and Their Support for Key Ethernet Service Features

Technology	Service Topology	Bandwidth	Granularity	Protection	QoS Classes	Overprovisioning	Network Segment
EoPON	Point to point	Up to 1 Gbps	Yes	No	Yes	Yes	Access
EoDSL	Point to point	< 100 Mbps limited by distance	Yes	No	Yes	Yes	Access
EoTDM	Point to point	Nx 1.5 Mbps Max ~ 40 Mbps	No	No	No resources are not shared	No	Access and core transport
EoS	Point to point	Up to 40 Gbps	No	Yes < 50 msec	No resources are not shared	No	Access and core transport
EoOTU (EoWDM)	Point to point	Up to 100 Gbps	Limited; min. 1Gbps	Yes < 50 msec	No resources are not shared	No	Access and core transport
EoHFC	Point to point	<100 Mbps Shared	Yes	No	Yes	Yes	Access
EoMPLS/PBB	Point to point, multipoint to multipoint	NA	Yes	Yes < 100- 500- 800 msec	Yes	Yes	Core Transport

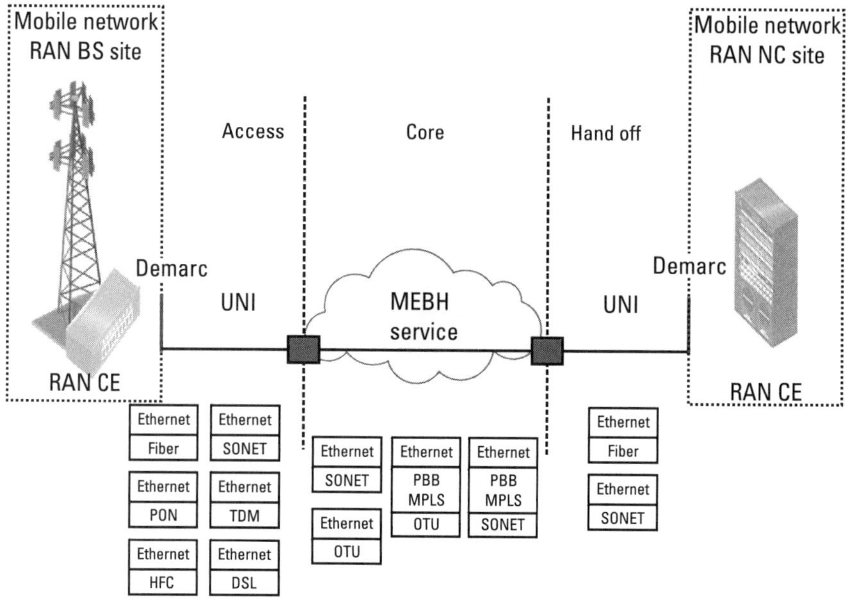

Figure 3.18 Ethernet delivery technologies data path.

which may in some circumstances provide a valid transport technology option (in the access segment of the service in particular). MPLS may be extended over to the access and handoff segments. The core transport may include more complex combinations of layers such as Ethernet over pseudowires [42], over MPLS, over Ethernet and over OTU. Does this matter? It does, as each technology has its limitations or advantages (i.e., certain technologies do not support some service attributes or support them only in a limited way. Knowing the context of the MEBH service will tell us more about the ability of MEBH service to support the service requirements than any other information. In addition, the overall Ethernet service properties are the result of the properties of the networking environment in which the Ethernet service is delivered. It is difficult to predict exactly how the properties of specific technologies underlying the service will affect the overall service. It is difficult, but it does not mean that it can, or should, be ignored.

3.3 Service Protection

3.3.1 Terminology

We begin with discussing the key concepts [43] describing resiliency and protection design and service. The terms are not standardized and may be used and interpreted in different ways. Thus, further discussion is beneficial to agree on what they mean.

3.3.1.1 Path

A path is a sequence of connected nodes and links with designated ingress and egress UNI and is capable of transferring traffic between ingress and egress CEs. Working path is the path used to forward service frames. Primary path is the preferred path for forwarding service frames between two or more UNIs. Backup path is a path that exists to carry service frames only if a failover event occurs on a primary path. Standby backup path is a backup path that is established prior to a failover event to protect a primary path. When a failover event is controlled by the customer, then the standby backup path will be a pre-established EVC.

3.3.1.2 Disjoint Path

For a service provider it means a pair of paths that do not share a common transport resources, such as links and nodes, other than ingress and egress UNIs. For a customer it means a pair of paths that do not share a common UNI.

3.3.1.3 Facilities

This is a physical resource in the transport network, such as a node, link, or path.

3.3.1.4 Protection

This is the architectural feature of a transport network that provides failure detection and failover from a primary service path to a backup service path or standby node when a failover event occurs. *Protection switching* is an action that redirects the traffic away from a working primary service path to a backup service path or standby node when a failover event occurs (e.g., a layer 2 switching deployed as the protection method). *Protection method* is a mechanism that performs protection switching. *Protection architecture* is a transport network architecture that provides link, node, path protection, and/or other facilities upon a failover event on a primary service path. Reliability is a somewhat similar term defined as the ability of the system to operate uninterrupted (i.e., survive failures) [44].

3.3.1.5 Resiliency

This is a qualitative description of the capacity of a transport network to withstand or recover from failed or degraded transport paths and facilities.

3.3.1.6 Redundancy

This is an architectural feature of a transport network that provides diverse facilities, such as standby nodes or standby paths over some or all of a primary service path.

3.3.1.7 Recovery

This is the action taken after a failover event whereby a node, link, or path is reinstated to its original state of performance.

3.3.1.8 Restoration

This is a state in which the primary service path has recovered from a failover event, but is not forwarding packets because the backup path remains the working path.

3.3.1.9 Reversion

This is the state of failover recovery in which the primary service path has become the working path so that it is forwarding packets. Protection switching may or may not support reversion. If supported, must occur after restoration.

3.3.1.10 Domain

This is a group of arbitrarily connected transport facilities, possibly with some common characteristics (same administration, same technology, same risk).

3.3.1.11 (Shared) Risk Domain

This is a group of transport facilities, which could be a node, a link, a building, or the combination of any of these sharing the same risk.

3.3.1.12 Shared Risk Group (SRG)

SRG is a set of facilities sharing a common physical resource, including links and nodes (i.e., sharing a common risk). SRG is a composite of SRLG, SRNG, SRDG [45]. *Shared risk link group* is a group of links sharing the same risk domain. *Shared risk node group* is a group of nodes sharing the same risk domain. Shared risk domain group is a group of transport facilities sharing the same risk domain.

3.3.1.13 Diversity

Diversity is the architectural feature of a transport network that supports disjoint shared risk groups (SRG) facilities for a primary service path. The concept of shared risk domains is illustrated in Figure 3.19 [46].

As an example we can differentiate several SRGs on the diagram. UNI-1, UNI-2, and UNI-3 are in SRNG with node N1, as the failure of the node N1 would interrupt the service to these UNIs. Node N1, link L1, and node

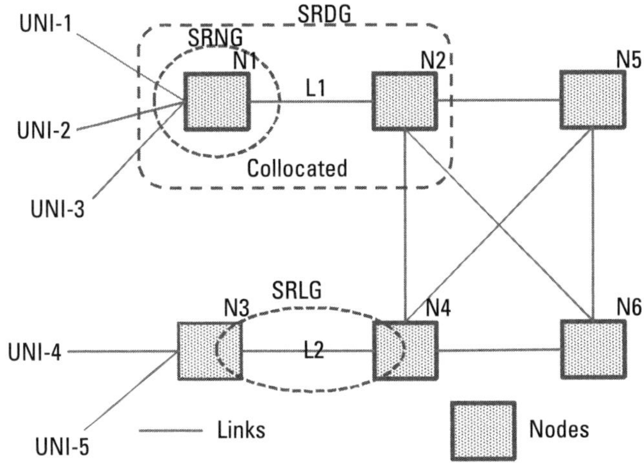

Figure 3.19 Exemplar shared risk domains.

N2 are in the SRDG, as they are placed in the same location and are subject to possibly the same fault conditions (e.g., flood or power outage). All services going through the link L2 (from UNI-4 and UNI-5) are in the same SRLG, as they share the same link, and the failure of this link would affect these services. Other SRGs in Figure 3.19 can also be identified.

3.3.2 Concepts

In the nutshell, the protected network must have extra facilities that can be counted on supporting service if some part of the network fails. The most elementary protection design is a linear protection schema protecting a link between two elements, in which two elements are connected over with two separate physical facilities, one of them active, another a stand-by service to protect the active one, as illustrated in Figure 3.20. In a case of failure, the traffic is switched from the active to standby.

In essence, the linear protection illustrates the generic concept of any protected service. In any variant of the protected design, the service has to have working and stand-by facilities, the failure detection mechanism responsible for detecting the failure of the working facility, and the switching mechanism that would switch the traffic between the working and stand-by facilities. Of course, in the real design we have complex protection configurations such as node protection, facilities protection, and virtualized facilities. Yet, it will not be an exaggeration to say that all of protected designs in essence resemble at some level of abstraction the protection configuration in Figure 3.20.

The resiliency of the network services and MEBH is a multilayer process. As we have mentioned in the previous sections, the Ethernet service is a layered

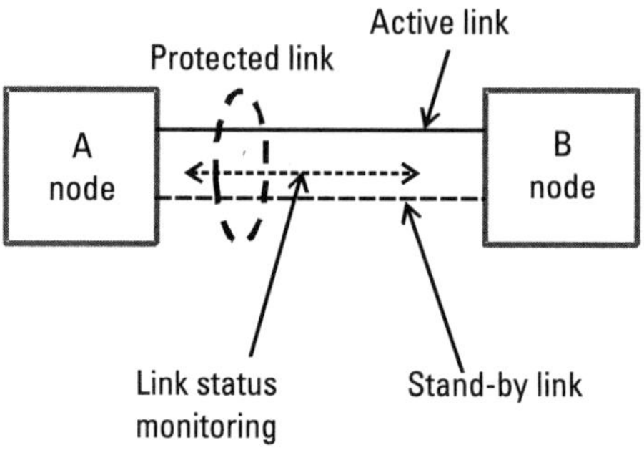

Figure 3.20 Link protection.

service. Always. Even in the simplest case of the native Ethernet transport, the Ethernet layer is riding over a layer 1 (i.e., the physical layer). It is because, every network service is in its essence a physical phenomenon. Thus, the resiliency of the Ethernet service is also a layered concept. This means that the resiliency of the Ethernet service is dependent on the resiliency of the layers below it. And, obviously, the resiliency of the layers above the Ethernet layer (that carry the services) depends on the resiliency of the Ethernet layer and all the layers below.

Higher layers cannot recover before lower layers recover. Thus, the total recovery time of a given layer is a sum of recovery times of lower layers. As well, each layer has so called timeout or hold-off time. The timeout is a time interval a services at the given layer can survive the lack of connectivity. For the service to be resilient or protected, its hold-off time should be longer than the recovery times of layers below them. Or, the recovery times of lower layers should be shorter (in sum) than the time-outs of the services on the layers above. If the lower layers have longer recovery times than time-outs of the layers above, the services at the higher layers cannot be protected. This concept is presented in Figure 3.21.

The layer recovery times (T_i) add up to the total recovery time. If, as on the diagram (B) of Figure 3.21 the hold-off time of the service at the layer x+1 is longer than the sum of recovery times of lower layers, then the service is protected. If, on the other hand, as on the diagram (A) the hold-off time of the service at the layer x+1 is shorter than the sum of recovery times of lower layers, then the service is not protected.

It is usual to compare any recovery times (of layers below layer 3) to the recovery time of SONET/SDH technology, which is around 50 milliseconds. Technologies of higher layers usually have longer recovery times—in a range of several hundred milliseconds to several seconds.

Some insights into the recovery process will foster better understanding of the protection processes. The recovery process is a series of events with time intervals between each event. Thus, the overall recovery time of the service is

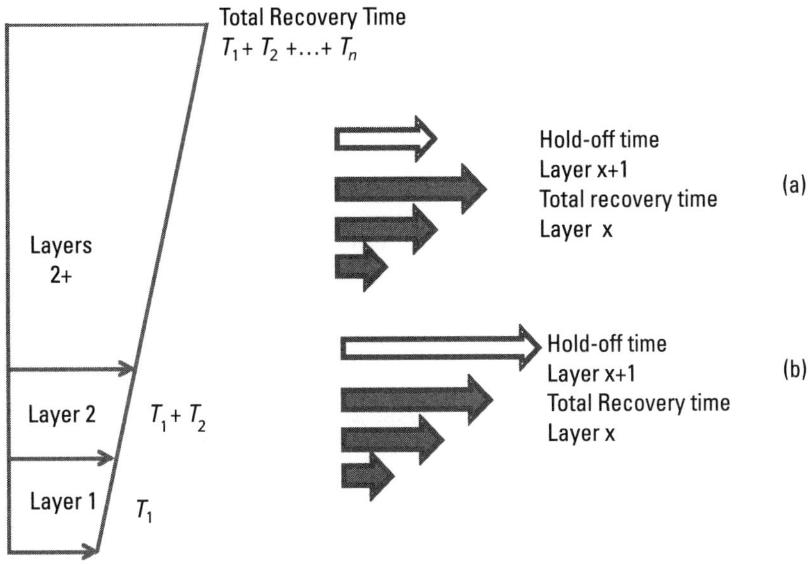

Figure 3.21 Protection in the layered service.

a sum of the component times of the recovery events. The recovery process is defined by the following events:

- Impairment event (Ie);
- Fault detection event (Fde);
- Hold-off event (Hoe);
- Connectivity restoration event (CRe);
- Service restoration event (Sre).

At some point in time (T) the network element or service fails (Ie)—this is termed an impairment event and the time of the event is Tie. The event is detected—this is termed a fault detection event (Fde). The time of this event is TFde. The control plane waits for certain time before the recovery process is initiated—this is called a hold-off event (Hoe). The hold-off event is terminated at time THoe. The recovery is completed and the connectivity is restored, which is a connectivity restoration event (CRe). This event happened at TCRe. After the connectivity is restored, the service may be restored, which is called service restoration event (SRe) at time TSRe [47]. Thus, the overall recovery time from failure to service recovery is the sum of all times listed above., and, one would add, at all layers carrying the service. This process is illustrated at Figure 3.22.

With some generalization, one may say that at every protocol layer the sequence of restoration events is similar.

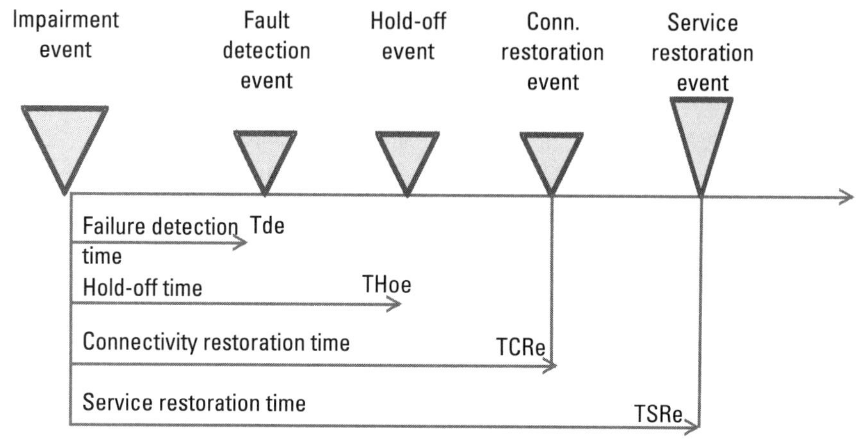

Figure 3.22 Restoration events and timing.

Why it is important to recognize multievent, multilayered nature of the restoration process? It helps the provider and the customer to understand the impact of particular elements of the restoration process on the service and fine-tune the design. What counts for the customer is not necessarily the restoration of the connectivity at the lower layers but the restoration of his service. Even if the connectivity between the elements providing the service can be restored quite quickly (TCRe), the service itself (TSRe) may take a long time to be up again.

In some architecture, even if the connectivity at the transport layers (usually layer 1 or 2) may be restored under several hundred milliseconds, the restoration of the service (at layers 2+) may take 10 or more minutes [48] if the service has been lost. Imagine losing wireless service over a large geographical area for this period of time during rush hour. It certainly looks bad for the wireless service. Thus, it may be the case that fine-tuning the fault detection and hold-over timers at the transport layers may save the provider long minutes of service blackouts, not to mention the poor PR and lost customers! How it is done? If the transport (or in general lower layers) can restore the layer-specific services within the hold-off time of the higher service layers, these layers may not declare the fault event and return to the service without interruption (i.e. without going into the full, long restoration process).

The protection design may be covering the UNI-to-UNI segment of the network, the UNI only, or any subsegment of the service. The better service protection is: less failures will affect the service and the service will recover in the shorter amount of time. To say it differently, protected service has high up time or availability.

Availability [49] of the service is expressed in the number of nines. Carrier grade networks are generally targeted to provide five nines (or higher) availability. Five nines means that out of some period of time, usually one year, the facility will be operation 99.999 percent of time. Table 3.18 provides the number of nines and the expected down time in a year, month, and week. One needs to understand that the number of nines expresses the long-term average behavior, not necessarily observable during a single reference period.

3.3.3 Estimating Availability

Availability is an elusive number and notoriously difficult to determine precisely. You could calculate the availability of service based on past experience. But this approach will tell us what did happen, not necessarily what will happen. In network design we want to be able to predict the resiliency of the design, before it is implemented, and compare different designs with respect to their reliability. The following method does just this. But a word of caution: as with any attempts to guess the future, this one is condemned to be as reliable as reading tea leaves. Yet it is "sanctioned" by the engineering praxis and imparts to the engineering estimates some aura of rational process. Availability is an estimate of the uptime. Five nines (0.99999) availability in a year means that the object with this availability will be functioning 0.99999 of a year, or, it will not be operational for five minutes in a year. This number can be associated with any element of the network. The networks can be at the level of abstraction presented as a construct of the parallel and serial objects connected together as in Figure 3.23.

Each object can be assigned its availability, interpreted as the probability of being in the operational state during the specific time. The overall end-to-end availability of such a construct may be estimated using a simple math as illustrated here.

Table 3.18
Availability Rating and Outage Times

Availability (Percent)	Downtime per Year	Downtime per Month*	Downtime per Week
90 percent ("one nine")	36.5 days	72 hours	16.8 hours
99 percent ("two nines")	3.65 days	7.20 hours	1.68 hours
99.5 percent	1.83 days	3.60 hours	50.4 minutes
99.8 percent	17.52 hours	86.23 minutes	20.16 minutes
99.9 percent ("three nines")	8.76 hours	43.2 minutes	10.1 minutes
99.95 percent	4.38 hours	21.56 minutes	5.04 minutes
99.99 percent ("four nines")	52.56 minutes	4.32 minutes	1.01 minutes
99.999 percent ("five nines")	5.26 minutes	25.9 seconds	6.05 seconds
99.9999 percent ("six nines")	31.5 seconds	2.59 seconds	0.605 seconds

* 30 days per month

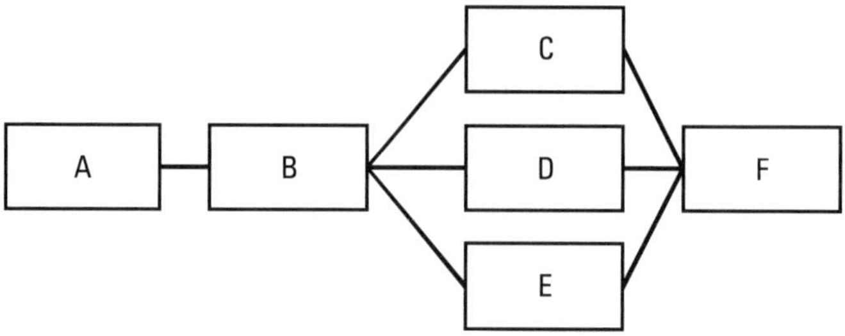

Figure 3.23 Abstract network model.

The objects A and B are connected together in series. Each object has availability of 0.9999. The availability of objects in series is estimated by the following formula:

$$A = \Pi A_i$$

where A_i is the availability of the i parallel component.

Thus, the availability of the A and B object is

$$A = 0.9999 * 0.9999 = 0.9998$$

The objects C, D, and E are connected together in parallel. Each object has availability of 0.99. The availability of objects in parallel is estimated by the following formula:

$$A = 1 - \pi[1 - A_i]$$

where A_i is the availability of the i component.

Thus, the availability of the objects C, D and E is

$$A = 1 - [1 - 0.99] * [1 - 0.99] * [1 - 0.99] = 0.999999$$

Thus, the summary availability of components can be substituted for the individual components; as a result, we have the series of components in series and the resultant availability is

$$A = 0.9998 * 0.999999 * 0.9999 = 0.99969$$

In principle every network can be reduced to such a calculation. Of course, the biggest problem is now how to interpret such a number in terms of the actual network behavior.

Availability [50] if not derived from the theoretical calculations [51] is estimated based on the past history of events. Two concepts are used here—that of mean time between failures (MTBF) and mean time to repair (MTTR). MTBF is the time the system or service has been operational [52]. MTTR is the average time it takes for the system or service to recover from failure. Availability for the specific system or service is calculated in the following way:

$$A = \text{MTBF}/(\text{MTBF} + \text{MTTR}) \times 100$$

In essence, the formula expresses the ratio of up time to total time elapsed. As with all statistical measures, this one is not the measure observable in a limited number of instances. It is an expected value over the long observation period. Thus, from time to time the highly reliable services and networks may fail (and they do) but do not lose their availability rating in the long run [53].

3.3.4 Protected Service Architectures

Many variants of network architecture can assure the reliability of the service. In general, the reliable service requires a resilient network. But it is not always the case. The highly resilient service may be implemented on the less resilient network. It means that the service with, for example, the five nines availability can be implemented on the network with the four nines availability rating. Following are some examples of resilient designs offering different levels of protection.

Figure 3.24 presents a design where the EVC has a unprotected handoff and an unprotected access facilities. The only protection offered here is the inherent protection offered by the MEN network. In other words, the resiliency of this design relies on the availability of the MEN network.

Table 3.19 summarizes the resiliency features of this design.

Figure 3.25 illustrates the protected handoff design. This design uses a single EVC but two handoff points. Only one handoff point is active at the

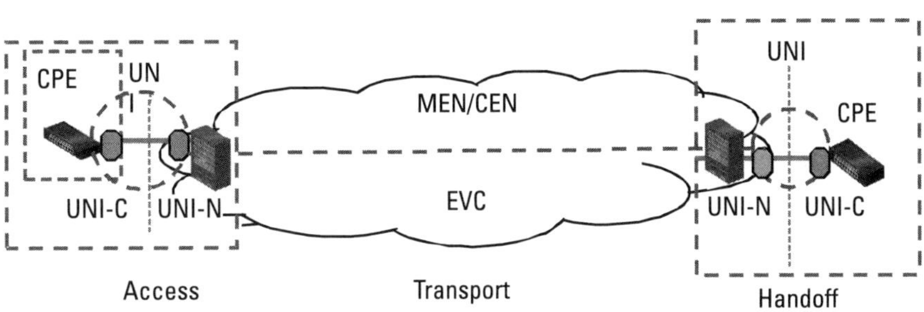

Figure 3.24 Single EVC, single handoff architecture.

Table 3.19
Failure Analysis—Single EVC Design

Network Segment	Protection	Failure Impact
Access	No protection in an access segment.	Access to the CPE at UNI affected; service interrupted.
Transport	Protection based on the MEN network design.	End-to-end service interrupted. If the MEN offers resiliency, the service may recover depending on the MEN resiliency design. The restoration time of several hundreds of milliseconds should suffice for most services with the exception of highly sensitive applications (e.g., emulated T1 (CES) pseudo wires).
Handoff	No protection in a handoff segment.	Handoff affected all EVCs at this UNI.
End to end	EVC is protected only from the failures in the MEN transport to the extent that the MEN network implements the resilient design.	Service is vulnerable to failures at handoff and access UNIs. MEN design may protect the service from transport failures. Such designs have usually less than four nines resiliency rating.

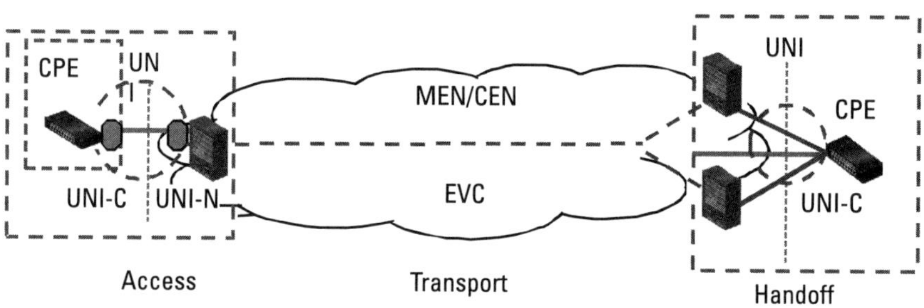

Figure 3.25 Single EVC with the protected handoff.

time. The other is used as a backup facility. Table 3.20 summarized features of this design.

Figure 3.26 illustrates a design that uses two EVCs and two handoff points. Thus, in principle the transport and handoff points are protected through independent paths. Of course, for this schema to work the two EVCs should not be in the same SRG facilities. Table 3.21 details the resiliency of this design.

Figure 3.27 presents the protected access, transport, and handoff segments. The service between the UNIs has two paths to traverse—one is active, and one is stand by. This is of course the most expensive protection mechanism. Table 3.22 details the resiliency of this design.

Figure 3.28 presents the variant of the design from Figure 3.27, where the protection facilities are provided by another carrier. Quite often this type of the design is used by customers demanding the ultimate failure resiliency. A design like this not only has to have independent dual paths, but it also has

Table 3.20
Dual Handoff Architecture

Network Segment	Protection	Failure Impact
Access	No protection in an access segment	Access to the CPE at UNI affected; service interrupted.
Transport	Protection based on the MEN network design	End-to-end service interrupted. If the MEN offers resiliency, the service may recover depending on the MEN resiliency design. The restoration time of several hundreds of milliseconds should suffice for most services with the exception of highly sensitive applications (e.g., emulated T1 (CES) pseudo wires).
Handoff	Dual handoff to the customer premises	If handoff fails, the service should switch to the standby handoff interface and the service should recover within the specified interrupt time The restoration time of several hundreds of milliseconds should suffice for most services with the exception of highly sensitive applications (e.g., emulated T1 (CES) pseudo wires).
End to end	EVC is protected from the failures in the MEN transport by the network protection mechanism and at the handoff by the dual handoff.	The overall service resiliency is improved as part of the service has parallel facilities. Such designs have usually around four nines resiliency rating with the resiliency of the handoff above four nines.

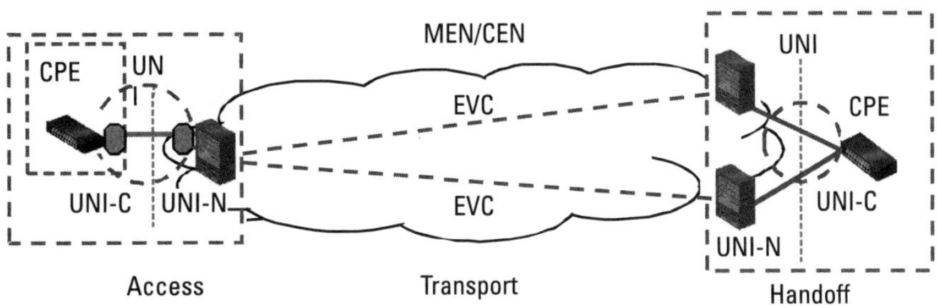

Figure 3.26 Two EVCs with two handoffs.

two different networks, assuming their independence, isolating events in one network from operations of the other one.

Of course, one may have dual access protection without the protection of the transport and handoff. Or, one may protect the access facilities but not handoff or transport. One may also protect part of the transport only. Thus, the possible combinations of protection design are many, and what is presented here are canonical cases only. The decision how to implement the protection design as it was discussed depends on which part of the network supporting the service is the most critical one to the service overall continuity [54].

Table 3.21
Two EVCS and Two Handoff Design

Network Segment	Protection	Failure Impact
Access	No protection in an access segment.	Access to the CPE at UNI affected; service interrupted.
Transport	Protection is achieved by dual transport paths—one active, one stand by.	End-to-end service interrupted. If the MEN offers resiliency, the service may recover depending on the MEN resiliency design The restoration time of several hundreds of milliseconds should suffice for most services with the exception of highly sensitive applications (e.g., emulated T1 (CES) pseudo wires).
Handoff	Dual handoff to the customer premises.	If handoff fails, the service should switch to the stand-by handoff interface, and the service should recover within the specified interrupt time. The restoration time of several hundreds of milliseconds should suffice for most services with the exception of highly sensitive applications (e.g., emulated T1 (CES) pseudo wires).
End to end	EVC is protected from the failures in the MEN transport and at the handoff by the dual handoff.	The service protection is improved considerably as the service has two parallel paths that are not in the same SRG (this is the assumption). The service should be protected from the failure in transport and handoff segments. Such designs have usually greater than four nines resiliency rating.

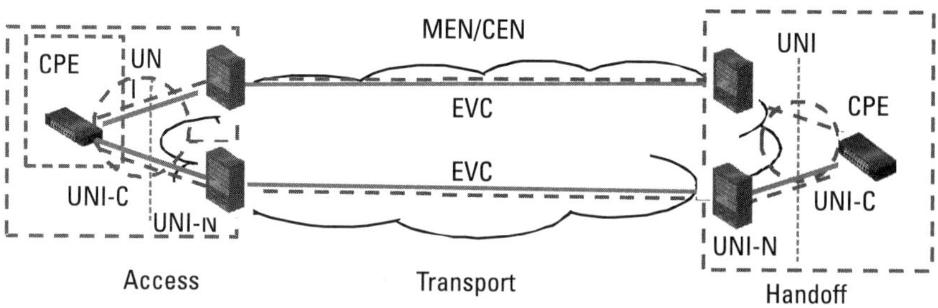

Figure 3.27 Two EVCs with dual access, dual handoff, and protected transport.

3.3.5 Implementation

The protection is expensive: it requires redundant facilities. Thus, in building protection in the network, one must exercise the sound judgment to strike a balance between redundancy, a required level of protection, and expected availability. One way to evaluate the effect of the protection design is to calculate the hypothetical cost of offering the protection to the number of megabits per second protected. For example, if the protection is built to protect the UNI interface with a 100-Mbps interface, the cost of the protection mechanism will be allocated to the UNI bandwidth of 100 Mbps. If the protection mechanism is built to protect a 10,000-Mbps UNI, the cost will be allocated to 10,000

Table 3.22
Two EVCs with Dual Access, Transport, and Handoff Facilities

Network Segment	Protection	Failure Impact
Access	Dual access.	Access to the CPE at UNI protected by the dual access facilities. If the stand-by and active paths do are not in SRG facilities, then the failure of one path should not affect the service. The restoration time of several hundreds of milliseconds should suffice for most services with the exception of highly sensitive applications (e.g., emulated T1 (CES) pseudo wires).
Transport	Protection is achieved by dual transport paths—one active, one stand by.	If the stand-by and active paths do are not in SRG facilities, then the failure of one path should not affect the service. The restoration time of several hundreds of milliseconds should suffice for most services with the exception of highly sensitive applications (e.g., emulated T1 (CES) pseudo wires).
Handoff	Dual handoff to the customer premises.	If handoff fails, the service should switch to the stand-by handoff interface and the service should recover within the specified interrupt time. The restoration time of several hundreds of milliseconds should suffice for most services with the exception of highly sensitive applications (e.g., emulated T1 (CES) pseudo wires).
End to end	EVC is protected from the failures in the MEN transport and at the handoff by the dual handoff.	The service is protected by two parallel paths, earning the high protection rating. Such systems may achieve five nines or six nines resiliency rating.

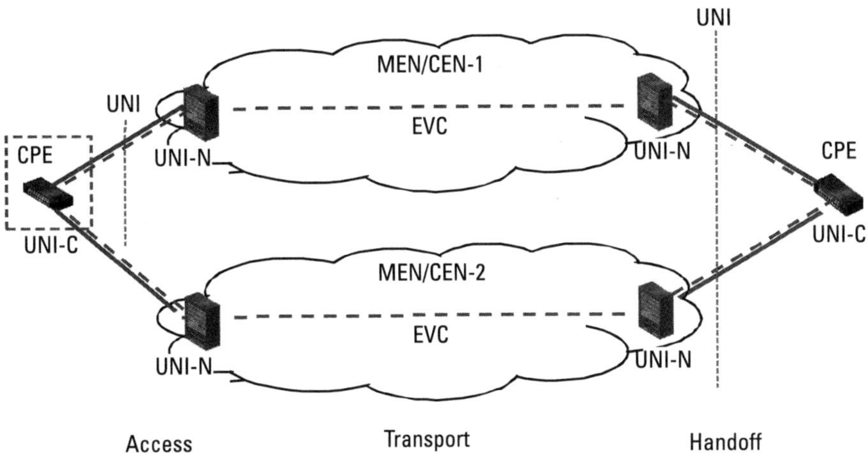

Figure 3.28 Two independent service providers.

Mbps. Thus, protecting access interfaces is expensive and rarely done. But the protection of the aggregated handoff interfaces is less expensive per bit, and that is why it is more frequently deployed. Other ways to evaluate the effectiveness of the protection is to compare number of EVCs projected or a number of UNIs protected. Of course, there is a common-sense interpretation of these metrics as well: it makes more sense to invest into the protection of facilities with many services riding on them rather than the protection of facilities with a small amount of services. This concept is illustrated in Figure 3.29. Protecting individual UNIs with a single EVC and 100-Mbps bandwidth is much less cost effective than protecting a handoff UNI with many EVCS and 10 Gbps of combined traffic. The hypothetical protection curve for Mbps and EVCs is provided in Figure 3.30.

3.3.6 Measuring Resiliency

Resiliency of every property of the service of network can be measured. Apart from the availability, other metrics can be used to characterize the operational characteristics of the network. The metrics [55] for this purpose are not widely used, but their understanding would be beneficial for the customer, as these metrics characterize (and provide insights into) the network protection design and its impact on the service. Of course, the metrics listed here are not an exhaustive list of metrics available for this purpose.

3.3.6.1 Failover Packet Loss

This is the amount of packet loss produced by a failover event until the failover completes, where the measurement begins when the last unimpaired packet is

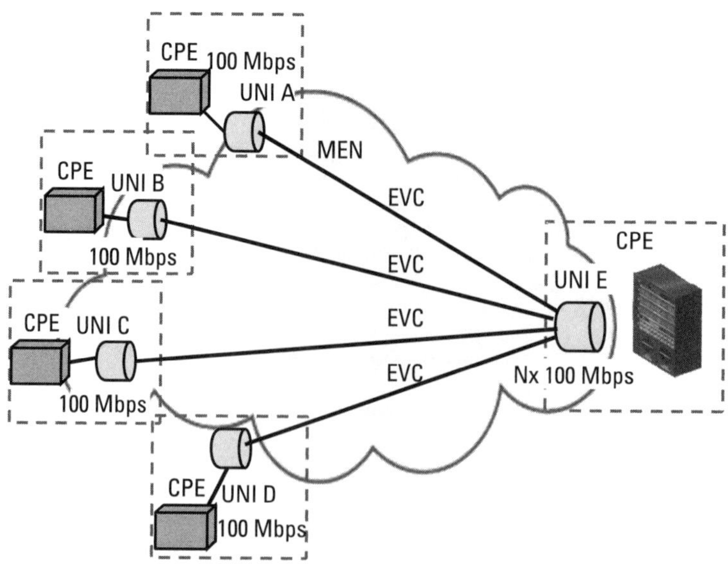

Figure 3.29 Different levels of traffic at single and aggregated UNIs.

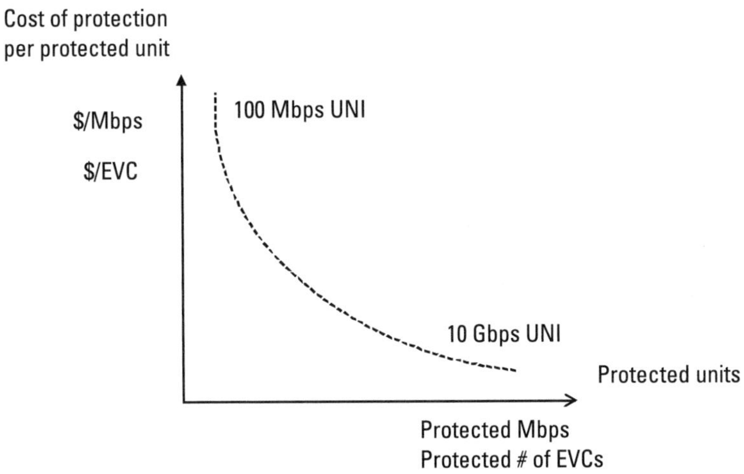

Figure 3.30 Hypothetical protection effectiveness curve.

received by the tester on the protected primary path and ends when the first unimpaired packet is received by the tester on the backup path.

3.3.6.2 Reversion Packet Loss

This is the amount of packet loss produced by reversion, where the measurement begins when the last unimpaired packet is received by the tester on the backup path and ends when the first unimpaired packet is received by the tester on the protected primary path .

3.3.6.3 Failover time

This is the amount of time it takes for the failover to complete so the backup path is established as a working path.

3.3.6.4 Reversion Time

This is the amount of time it takes for reversion to complete so that the primary path is restored as the working path.

3.3.6.5 Additive Backup Delay

This is the amount of increased forwarding delay resulting from data traffic traversing the backup path instead of the primary path. Additive backup delay is calculated using the formula shown here:

$$\text{Additive Backup Delay} = \underset{(\text{Backup Path})}{\text{Forwarding Delay}} - \underset{(\text{Primary Path})}{\text{Forwarding Delay}}$$

While these metrics maybe difficult to measure in real services, one may use these concepts to quantify the properties of the network. For example, one may establish as part of the service specification a maximum failover time and reversion time required for the service on failure event. One may also explicitly request that the additive backup delay is bound by the same limits as a delay time on the primary path.

3.4 Quality of Service

The quality of service (QoS) mechanisms are designed to manage transport of service frames through the network assuring some level of consistency in treating of packets in case of congestion (observe that fully provisioned networks such as SONET do not have similar QoS mechanisms, as they do not need to resolve congestion conditions). Depending on the class of service (CoS) assigned to service frames, these mechanisms may pass certain frames ahead of others, delay some of them, or drop them. In general QoS mechanisms enforce service parameters grouped under the concept of QoS service classes. Service frames with a so called high QoS class get better performance characteristics than do frames with lower CoS classes. The specific CoS class of a frame may be signaled in the network using the PCP field in the Ethernet VLAN tag (in-band signaling) or imposed by the out-of-band provisioning. The QoS characteristics of the service are organized into the QoS profiles or QoS policies. The QoS profile includes bandwidth parameters and flow performance characteristics per class of traffic. The following sections describe elements that constitute the QoS profiles as well as mechanisms that enforce QoS policies [56].

3.4.1 Bandwidth Concepts

The bandwidth or transmission rate is the number of traffic units (bits, bytes, octets, frames, or packets) per second transmitted over the interface. The transmission rate may be expressed in bits per second, frames or packets per second, or octets per second. Depending on which transmitted bits are counted toward the rate, we can define protocol transmission rate (counting bits of the specific protocol only), physical transmission rate (counting all the bits passing through the interface), or service transmission rate (counting only the bits that are part of the service). Bits per second can be used to characterize both physical media and services. Frames are most often used to characterize services. The frame used to characterize the transmission rate may include all or some aspects of the protocol the service frame belongs to. In the Ethernet services we define the concept of the service frame. The service frame includes all the fields in the Ethernet II frame from the first bit of the MAC DA address field to the last bit of the FCS field, but it excludes the control fields PRE, SDF, and IFG. The control fields are referred to as Ethernet protocol overhead. For the same bit rate, one will get different frame rates depending on whether one counts frames with or without the overhead.

Frames are transmitted with certain rate of frames per second. The maximum long-term average rate of transmission for which the delivery of frames is guaranteed is called committed information rate (CIR) and is characterized by bits per second. CIR approximates the transmission in which frames are sent with relatively equal interframe intervals as illustrated in Figure 3.31.

Figure 3.31 illustrates the transmission rate of 8 frames per second (fps) of 100 bits per frame, or 800 bits per second, or 100 octets per second. The same rate of transmission may be achieved when the frames are transmitted over shorter than 1-second time frames followed by the periods of no transmission, as illustrated in Figure 3.32.

In this case, all frames are transmitted in a fraction of a second followed by the period of no transmission. Such a transmission is called bursting (opposed to the uniform transmission from the previous example) or back-to-back transmission. The parameter that defines the ability of the interface to transmit burst of traffic is called committed burst size (CBS). CBS is measured in bytes.

3.4.2 Bandwidth Enforcement

3.4.2.1 Policing

The bandwidth is defined by CIR, CBS, EIR, and EBS parameters. We have explained already CIR as a long-term average guaranteed bit rate and CBS as a number of bytes that can be transported back to back without violating CIR. The excess information rate (EIR) and the corresponding excess burst size (EBS)

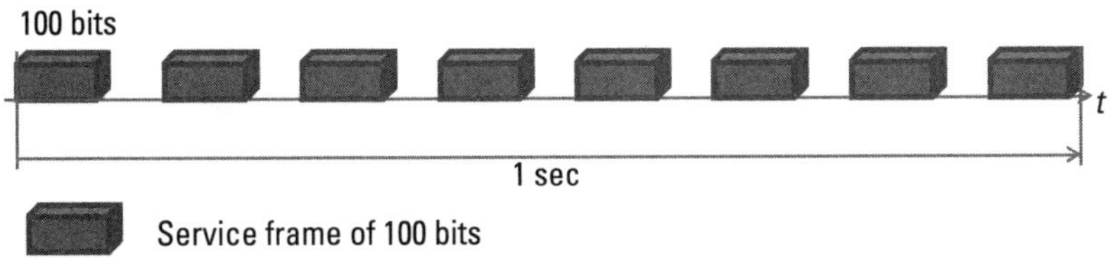

Figure 3.31 Transport of service frames at regular intervals.

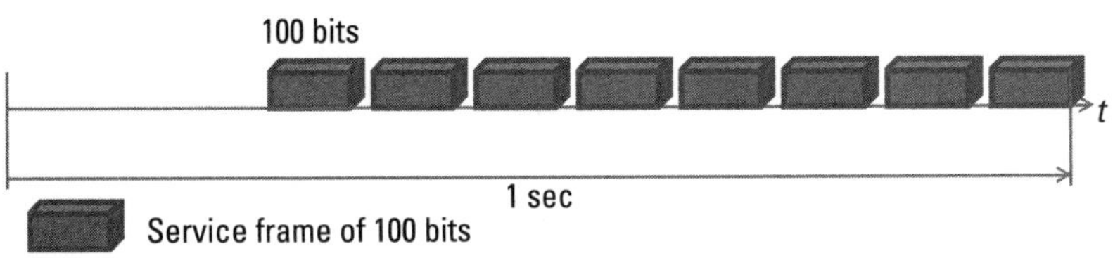

Figure 3.32 Back-to-back transport of service frames.

are parameters similar to CIR and CBS. They allow the CIR rate to increase to EIR and CBS to EBS at times when no other traffic is present on the interface. The policers with EIR/EBS parameters can support sustained bursts rather than only limited bursts supported by CBS.

If the QOS management system of the network device is color aware, the traffic that is above CIR but below the EIR rate is called and marked "yellow." Traffic that is below CIR is called "green." And traffic above the EIR rate is called "red" [57] and may or may not be dropped. In a case of congestion of transmission queues, the yellow traffic will be dropped before the green traffic using rules called RED (or WRED) [58]. The color-aware policers, in addition to marking traffic as indicated, are also aware of the color of incoming traffic and provide color-specific treatment. The packets marked yellow when coming to the policers are processed at EIR/EBS rate. In policers that are colorblind, all incoming packets are treated the same way.

The operation problems with the CBS size and traffic bursting are illustrated in Figure 3.33 [59]. The upper graph illustrates the arrival of service frames. The graph at the middle shows the status of the CBS token bucket. The graph at the bottom shows the refresh events. If frames arrive with the spacing matching the CBS (token bucket) refresh rate, there is enough token in the CBS token bucket to service the frames and maintain the CIR. If the frames arrive back to back, the CBS does not have enough tokens to service all arriving frames, before the CBS token bucket is refreshed and frames are dropped

The enforcement of the CIR/EIR rates is done using policers. Figure 3.34 illustrates the single-rate policer mechanism. The single-rate policer enforces

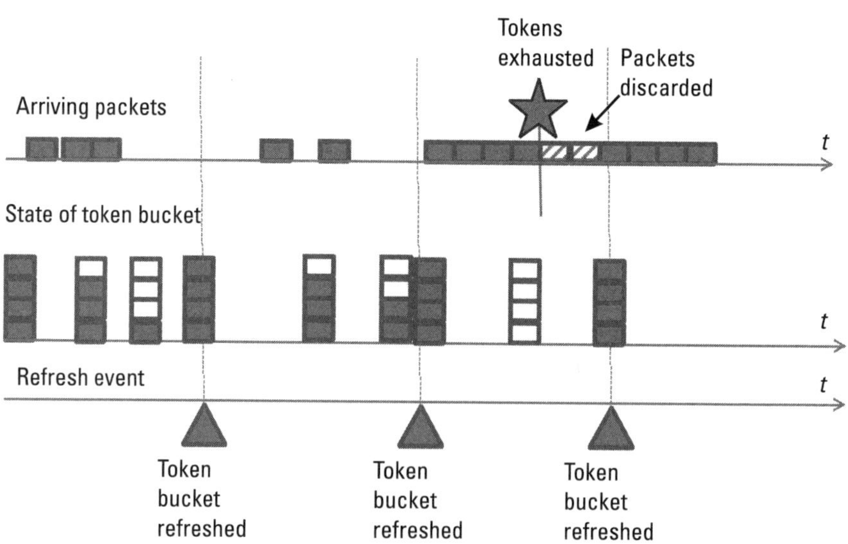

Figure 3.33 Servicing of bursty traffic.

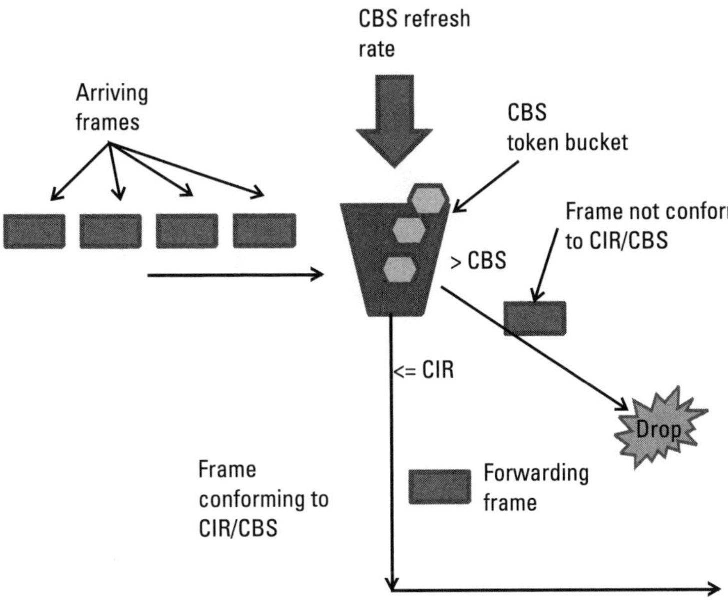

Figure 3.34 CIR/CBS policer (single rate).

the CIR/CBS parameters only. The algorithm implementing the single-rate policer is called a single-token bucket algorithm. The token bucket allocated for the policer has the size equal to CBS. The "bucket" is refreshed with new bytes or tokens [60] at the CIR rate. The token bucket can be refreshed continuously (using specific quanta) or at the specific time intervals. The method of refreshing the token bucket will affect how the policer will process traffic.

If the frame of size N is present at the interface and the number of tokens in the bucket is greater or equal to N, the frame is transmitted; otherwise, it is dropped.

The EIR/EBS parameters are enforced using the three-color two-rate policer illustrated in Figure 3.35. This algorithm in addition to the CBS bucket has the EBS bucket refreshed with the EIR rate. The CIR and EIR rate may be different. Continuing the previous example, if the CBS bucket does not have enough tokens to allow the frame to pass, the frame is marked yellow and the state of tokens in the EBS bucket is checked. If the amount of tokens in the EBS bucket is less or equal to the frame size, the frame is passed. Otherwise, the frame is marked red and may be dropped.

The CBS size must reflect the CIR and the refresh rate of the CBS token buffer. If the refresh rate is T_f in a fraction of a second, to sustain the CIR rate the minimum CBS buffer is expressed by the formula

$$\text{CBS} = \text{CIR} * T_f$$

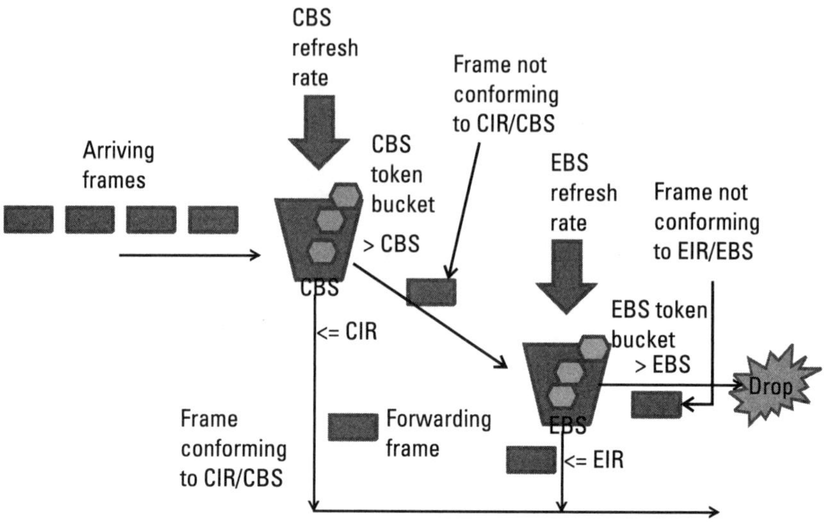

Figure 3.35 CIR/EIR policer (three-color two-rate policer).

Thus, for the CIR of 50 Mbps and the T_f of 1/10,000 seconds, the minimum CBS is 50,000 bits or 625 bytes [61]. CBS is usually defined considering the specific switch configuration requirements and the MTU size.

All frames that pass through the policer are queued for egress. If the queue fill rate exceeds a certain level of q queue fill (percent of queue occupancy), frames are dropped from the queue. Frames may be dropped randomly, in the reverse sequence of arrival, or using a congestion-avoidance algorithm.

3.4.2.2 Congestion Avoidance

Congestion avoidance is a method for active management of queues holding packets before they are transmitted. The congestion avoidance algorithm calculates the average queue size and then based on the rules drops the packets decongesting the queue. The packets may be dropped from the queue in many ways. The simplest way is to drop packets at the end of the queue (i.e., those that arrived last). This is called a tail-drop algorithm. However, nondiscriminatory packet dropping may have an adverse effect on certain higher-level protocols. A more sophisticated method uses a measure of the congestion and a type of packet to select packets singled out for dropping. The algorithm like this is called RED. This type of an algorithm is presented in Figure 3.36.

The graph represents the behavior of the packet discard algorithm. When the queue occupancy (X-axis) is growing from 0 percent to A percent, the probability of dropping a packet is increasing gradually from 0 to k. In the region of the queue occupancy from 0 percent to A percent, time packets are dropped from the queue at random with the probability defined by the graph. At point

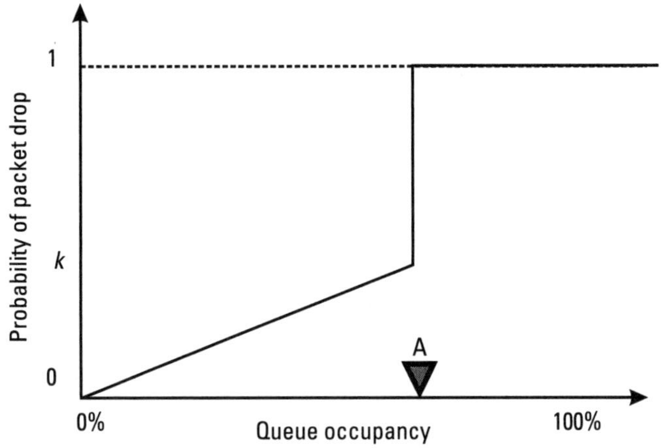

Figure 3.36 A single drop profile.

A percent, all packets are dropped as the probability of packet dropping is 100 percent. A more sophisticated packet-dropping mechanism is presented in Figure 3.37.

In this method the queue decongestion algorithm is sensitive to types of packets. Based on the type of packets, different probabilities of dropping, depending on the queue congestion, are used. The packets may be classified into different drop profiles based on their class of service, their color (if the device implements color-aware policing—yellow or green), or a combination of factors. Such management of congestion allows for the more selective traffic management. The congestion management using this method is called weighted RED (WRED).

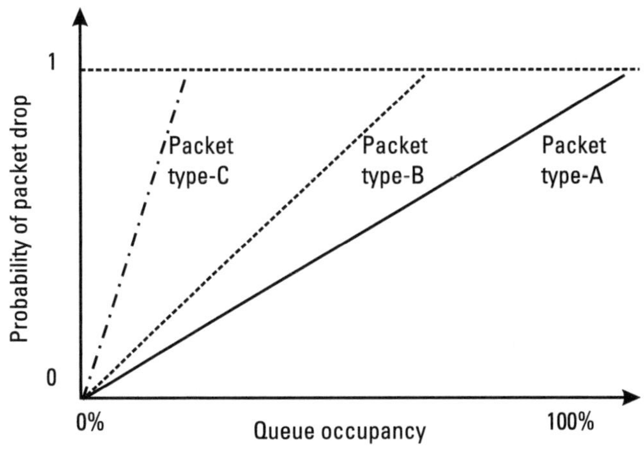

Figure 3.37 Per packet type drop profile.

3.4.2.3 Policing Versus Shaping

Policing of traffic means imposing certain average, long-term CIR rates on the packet flows. This is done at the risk of dropping frames that do not conform to the traffic specifications (CIR, CBS, EIR, or EBS). Another way to create a traffic flow that conforms to bandwidth parameters is to shape the traffic. Quite often traffic shaping subsumes any mechanisms used to create packet flows with specific characteristics. Here, traffic shaping denotes mechanisms that allow us to delay packets to create the required CIR and CBS parameters. The effect of packet shaping versus packet policing is visible on the graphs in Figure 3.38 and Figure 3.39. With the traffic policing any excess traffic above the specific CIR rate is dropped. This is symbolized in Figure 3.38 with shaded areas on the

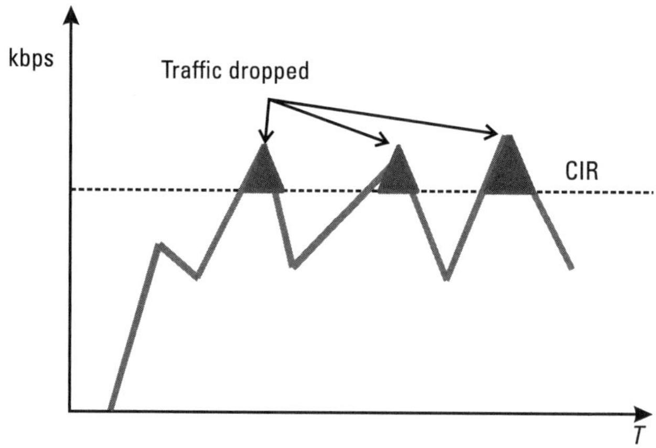

Figure 3.38 Profile of policed traffic for CIR.

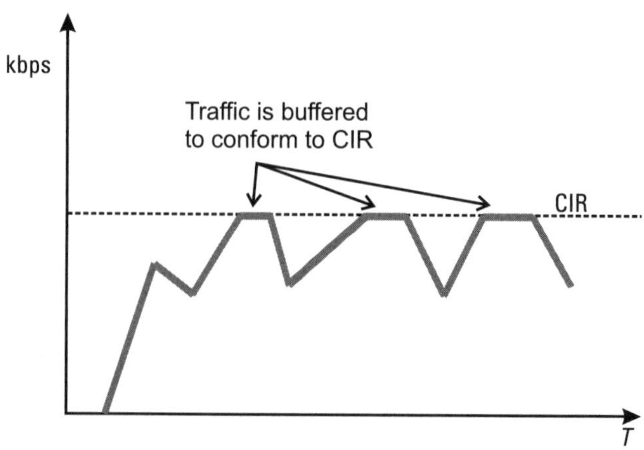

Figure 3.39 Profile of shaped traffic to CIR.

graph. Obviously, such a treatment of packet flows will result in large packet losses expressed in high frame drop ratio (see SLA metrics) and may limit the traffic flow below the CIR rates, due to the congestion avoidance mechanism in TCP protocols. If, instead of traffic policing, the service uses traffic shaping, packets that exceed the CIR are retained within the permissible queue size and then sent out. The resulting packet flow conforms to the CIR rate without incurring the frame losses. However, all this is achieved at the cost of increasing the delay and delay related impairments (interface delay, frame delay range).

3.4.3 Bandwidth Profiles

The bandwidth (BW) profile is an organizing concept that refers to parameters imposed on the traffic flow—CIR,CBS, EIR, and EBS. The BW profile may be applied to the UNI, to the EVC, to the COS, or to a combination of these objects. Of course, one may imagine a system without any BW profiles. This is quite common in the simple network interface devices. In such cases, the physical capacity of the port constituting the UNI defines the traffic flow characteristics.

3.4.3.1 BW per UNI

If the BW profile is applied only to UNI as illustrated in Figure 3.40, the BW parameters are applicable to all traffic ingressing UNI. All frames are subjected to the same set of CIR, CBS, and optionally EIR and EBS.

Parameters such as BW profiles are used for all-to-one port services. Or, they may be used as components of the more complex hierarchical profiles (i.e., BW profiles in which parameters are applied in sequential order to different elements of the traffic flow).

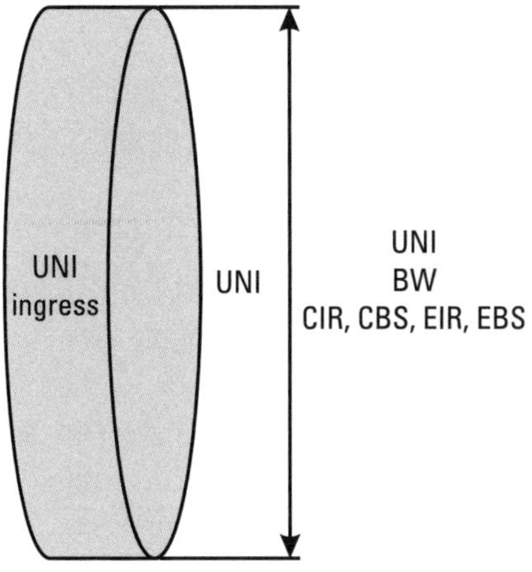

Figure 3.40 Bandwidth profile applied to UNI only.

3.4.3.2 BW per EVC

The bandwidth profile may also be applied to the EVCs on the UNI, as in Figure 3.41. In this case each EVC present on the UNI has a separate set of parameters CIR, CB, EIR, EBS, and each frame is managed according to the EVC it belongs to. Of course, there is an implicit UNI limitation, as UNI is a physical resource with the limited capacity to handle traffic.

3.4.3.3 BW per CoS

One may implement the bandwidth profile applied to the specific class of traffic, as in Figure 3.42. In this construct the BW parameters are applied to the flows differentiated by the specific COS class. It is rare that such BW profiles are used on their own rather than as a part of more complex, hierarchical schema.

3.4.3.4 BW per UNI, EVC, and CoS

In addition to BW profiles per UNI, EVC, or CoS, we may have BW profiles that combine the UNI, EVC, or CoS BW profiles into more complex designs. These profiles are called hierarchical, as they impose multiple, nested, levels of traffic parameters.

Figure 3.43 presents the concept of the BW profile applied to UNI and EVCs. In this case each EVC has a separate BW profile. In addition a combined traffic that ingresses UNI is a subject to the UNI BW profile. As a result the traffic on each EVC cannot exceed a certain limit, and traffic combined from the EVCs cannot exceed the UNI traffic profile parameters.

For example, two EVCs are provisioned on the UNI. Each EVC supports 50 Mbps of CIR. However, the UNI has also a CIR of 50 Mbps. Thus, the combined traffic over this UNI cannot exceed CIR of 50 Mbps on both EVCs with the effect that only one EVC may transmit the traffic at the full CIR rate.

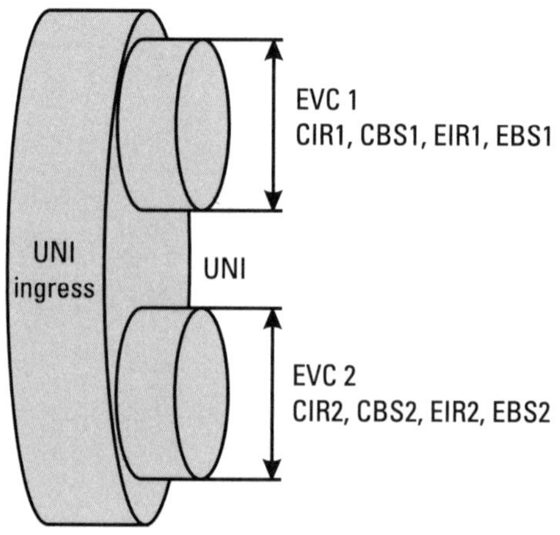

Figure 3.41 Bandwidth profile per EVC.

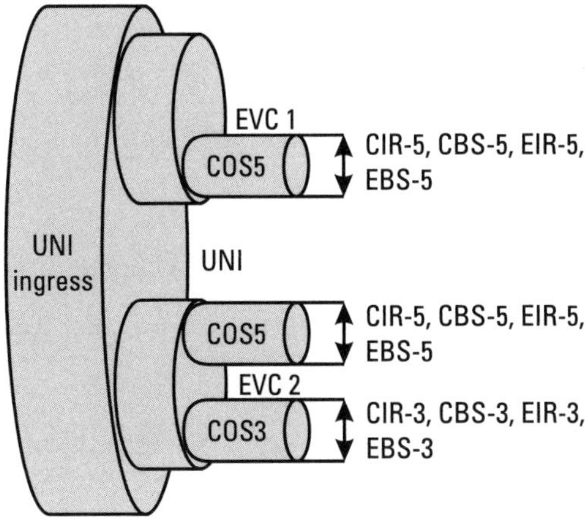

Figure 3.42 Bandwidth profile per COS.

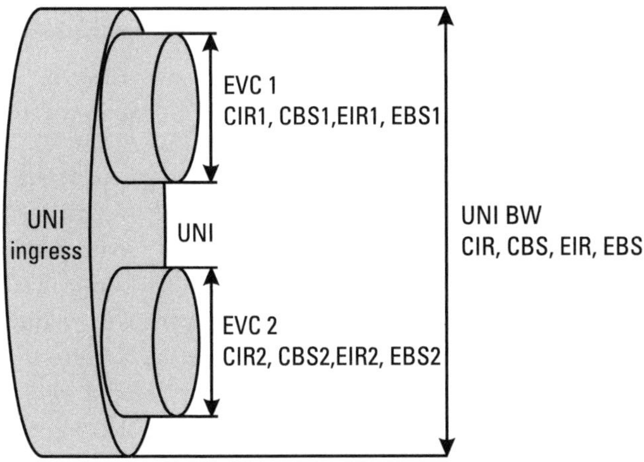

Figure 3.43 BW profile per UNI and per EVC.

Another example of the complex traffic profile is illustrated in Figure 3.44. In this example there is a CoS profile for each EVC. In addition, there is an overall COS profile per UNI limiting the amount the specific CoS can carry per the UNI. Such a design could be used in the scenario where each EVC has allocated 40 Mbps of H class of traffic. However, the combined H class traffic on all EVCs cannot exceed CIR of 80 Mbps. Many other combinations of hierarchical use of BW profiles are possible and are only limited by the specific implementation.

The bandwidth profiles can be applied to the ingress frames (ingres BW profile) or to the egress frames (egress BW profiles). The mechanisms used to

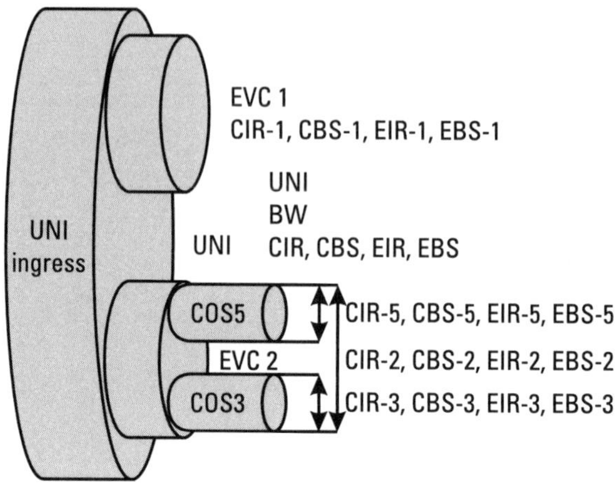

Figure 3.44 BW profile per EVC and CoS.

realize these profiles are essentially the same. However, the implementations for ingress versus egress may differ in details.

3.4.4 Estimating Throughput

The throughput (throughput rate) denotes the amount of traffic in bits per second or frames per second that can be transported over the UNI or EVC. Other terms that are used to denote throughput rate are bit rate, bandwidth, traffic capacity, and channel capacity. One may also encounter a term goodput denoting an amount of useful information (expressing, for example, a size of the file) that can be transferred in the unit of time. In contrast, a raw bit rate (called also line rate) expresses the number of bits delivered in a unit of time. Some terms are more likely to be used in the context of the specific technology—for example, the term channel capacity would denote throughput of the physical or logical circuit.

In Ethernet service the throughput rate may refer to the Ethernet frames with or without the control fields (IFG, SFD, PRE), to the service frames [62] only (excluding control frames), or the total amount (physical number of bits) of traffic that can be handled by UNI or EVC.

In most cases what is of interest to the customer is the throughput rate expressing the amount of user data (meaning service frames only) that could be transferred. The following definitions of throughput have not been approved by any standard organization but they are quite often used to denote different throughput measurements of Ethernet services.

Nominal throughput is the capacity of the transmission element (node, port, and line) expressed in terms of the transmitted protocol units. Quite often it is equated to the line rate. The line rate for Ethernet technology includes all transmitted payload bits and the interframe gap (IFG), preamble (PRE), and start frame delimiter (SFD) (see Table 3.23)—all together 20 bytes per each Ethernet frame. These fields are called an overhead.

All physical ports have nominal throughput capacities associated with them; for example, a 100-base T port has a nominal throughput or line rate of 100 Mbps. Some denote this throughput as a protocol throughput.

Effective throughput (called information rate) is the capacity to transport the service frames (excluding the protocol-related fields) only. The fields in the Ethernet frame that counts toward the Ethernet effective throughput are the bits contained within the start of the destination address through the end of the frame check sum (FCS) bits and exclude control fields. For Ethernet technology, three main factors influence the *effective throughput*, or the number of bits of user data that can be successfully transmitted via Ethernet frames:

1. Nominal throughput capacity;
2. Protocol overhead;
3. Frame size.

The dependencies between throughput, overhead, and frame sizes are illustrated in Table 3.24 for 50 Mbps of the nominal throughput.

Table 3.23
Ethernet Frame Overhead

Parameter	Symbol	Value
Interframe gap + start frame delimiter	IFG +SDF	12 B
Preamble	PRE	8 B
Overhead total	OH	20 B

Table 3.24
Effective Throughput Rates for Different Packet Sizes

Frame Size (FS) (Bytes)	Effective Throughput Rate (ETR) (Mbps)	Effective Throughput Rate (ETR) (pps)	Nominal Throughput Rate (NTR) (Mbps)
64	38.10	74414.04	50
256	46.38	22646.48	50
512	48.12	11748.04	50
1522	49.35	4053.5	50
2000	49.50	3093.7	50
9600	49.90	649.7	50

The effective throughput rate for the nominal throughput rate was calculated as follows:

$$ETR = (FS/(FS + OH)) * NTR$$

It is critical then for the understanding of the throughput to consider the method with which throughout measurement is derived, including the consideration of the type of frame, any overhead, and the specific layer—layer 1 (line rate), layer 2 (Ethernet with our without OH), layer 3 (IP rate), layer 4 (UDP rate), or even higher layers.

Figure 3.45 illustrates the throughout concepts depending on which protocol layer is considered [63]. The line rate is a physical interface rate. The Ethernet frame rate is the rate based on the Ethernet service frame (no overhead). The IP rate is a rate based on the IP payload carried by the Ethernet frame. The UDP payload rate is a rate of the UDP protocol payload. In general, as the frame size increases and the overhead decreases (relatively) rates asymptotically approach (but never reach) the line rate of the interface. The smaller the frame size is, the larger is the gap between the different throughput rates. The gap is decreasing as the frame size increases, but the rates never equal each other. The TCP [64] rate is a separate issue, as it is dependent on factors other than overhead. One may estimate the TCP throughput using the formula

$$TCP\_Throughput = TCP\_W/RT\_FTD$$

where:

TCP_W is a size of the TCP receive window.

RT _FTD is the round trip frame transfer delay.

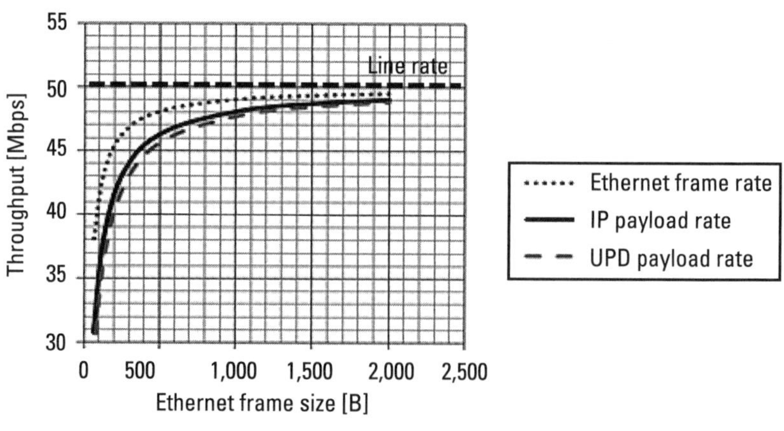

Figure 3.45 Throughput versus different rates on 100-Mbps interface.

The dependency between the CBS size and the TCP throughput should be clear. If frames within the TCP receive window are sent as burst they may exceed the CBS size, even if on the average they are sent within the CIR rate. Such frames would be dropped, forcing the TCP to lower the receive window size and retransmit frames. Under such conditions the service would never achieve the contracted CIR rate, even if the QoS configuration would support the requested CIR [65].

3.4.5 QoS—End-to-End View

It may be helpful at this point to provide a picture of how all the elements that mange the QoS policies fit together in the layer 2 (Ethernet) switch. The high-level view of the QoS components is presented in Figure 3.46.

The QoS design has two components blocks—ingress and egress. The ingress QoS component applies the QoS polices to the ingress frames coming from the ingress interface. At the ingres QoS component, packets are classified according to the marking policies and policed according to the bandwidth enforcement rules specified for the given UNI or EVC. From the ingress, interface packets are sent to the egress component that applies egress QoS policies and sends frames into the egress interface. At the egress, QoS components packets are classified according to the CoS classification rules. Packets are placed into the specific, class queues and then policed and or shaped.

This process is illustrated in more detail in Figure 3.47.

Frames coming from the ingress interface are classified and policed, or, to say it differently, the bandwidth policies are applied to the incoming frames. Frames are then queued according to their class of services. Queues are assigned specific CoS classes. Usually there are one or two queues called high-priority queues that are reserved for packets with the highest class of traffic. Other queues are designed for packets with lower classes of CoS. The congestion

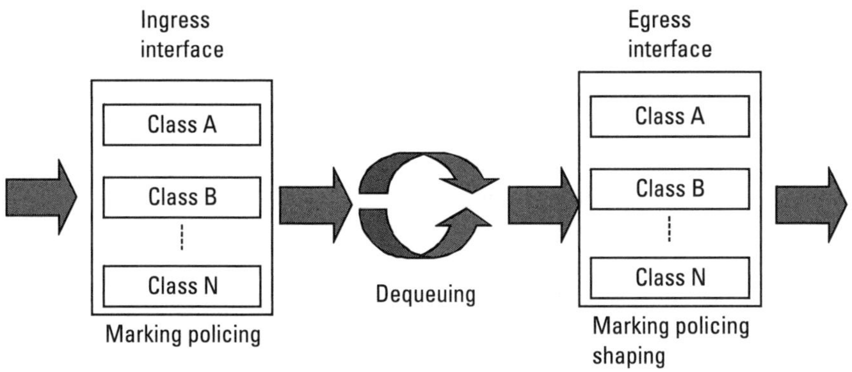

Figure 3.46 High-level QoS design.

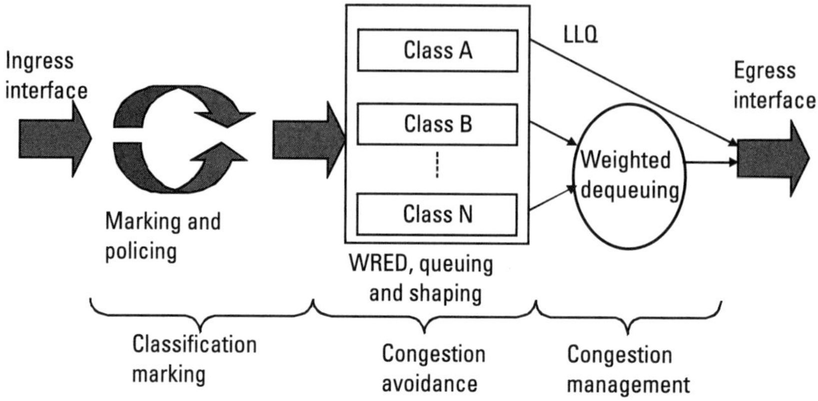

Figure 3.47 QoS polices mechanism.

in queues is controlled with a WRED algorithm. The dequeuing algorithm forwards the frames according to the priorities into the egress interface. The highest-quality queues have a separate dequeuing algorithm giving packets in these queues an absolute forwarding priority (low latency queueing, or LLQ, bypasses other queues). Other queues are serviced based on some version of the weightier round robin (WRR) algorithm. The shaping may be applied to the egress frames to assure that the traffic leaving the interface has specific flow properties (CIR, CBS), or no excessive bursts that could impair the quality of service.

3.4.6 CoS Classes

3.4.6.1 Marking Classes

We need to pause for a moment and explain the concept of the CoS. We already said that QoS defines certain attributes assigned to the traffic flows that impose a certain behavior on the traffic. In Ethernet service network flows with different behavior have assigned different designations called CoS class. This assignment to, or designation of, CoS class may be through the EVC identifier, port, or markings on the header or data portion of the Ethernet frame. If the service frame is tagged, the VLAN header has a three-bit field called priority code point (PCP). The PCP field can have values of 0 to 7. PCP bits define the specific CoS class to which the packet belongs. Customarily, PCP=0 designates the lowest COS class. The PCP=7 designates the highest. Thus, in all eight classes of traffic can be defined with the PCP bits. Of course, the frame with PCP coding is not any different from the frame without the coding. PCP coding only determines how the network devices that recognize this coding will treat the frame.

Other means of marking CoS class of packets in the Ethernet services also exist. For example, one may use the IP DSCP field on the IP payload in the Ethernet packet to designate the class of traffic. However, such a method requires first that the Ethernet frame carries the IP packet and that the UNI is able to read the DSCP field. The marking of the DSCP field with the CoS class is the only way to mark the packet CoS class if it is not tagged with VLAN tag.

3.4.6.2 CoS Classes

Classes of traffic define a specific behavior of the packets within this class. Each traffic class can be used for the specific service for which the characteristics of the class are best suited. As the network devices do not necessarily recognize all eight classes of traffic, quite often only three or four classes of traffic are used. Table 3.25 provides an example of several possible CoS designs and different services that are best served by the specific CoS flows.

In the CE service most often four traffic classes are differentiated. These classes are denoted as high plus, high, medium, and low. However, due to the limitation of the networking devices, four classes may have to be reduced three or two CoS classes. Table 3.25 presents one of possible options to implement mapping of services to different numbers of service classes.

3.5 Service Performance

3.5.1 Metrics

Each CoS class in addition to BW profiles has associated performance objectives expressed as performance metrics targets. Metrics define the characteristics of the packet transmission that are critical in delivery of services or important to the understanding of the network and service state. Metrics most often used

Table 3.25
Exemplary Designations of CoS

	Generic Traffic Classes Mapping to CoS Names			
CoS Names	**4 CoS classes**	**3 CoS classes**	**2 CoS classes**	**2 CoS classes**
Very high (H+)	Synchronization			
High (H)	Conversational, signaling and control	Synchronization, conversational, signaling, and control	Synchronization, conversational, signaling, control, and streaming media	Synchronization, conversational, signaling, control, and streaming media
Medium (M)	Streaming media	Streaming media	-	Interactive and background
Low (L)	Interactive and background	Interactive and background	Interactive and background	

define the transmission time and variation in the transmission time of packet, packet loss, and availability of the service. The packet transmission time is often referred to as delay or latency. The packet loss is referred to as a frame or packet loss or a delivery ratio (which is a complement of the packet loss). The variation in the transmission time is referred to as delay or latency variation or jitter. One must be careful using these apparent synonyms, as they refer to different concepts.

Metrics, as quantities describing a physical system, have bounds. The transmission delay cannot be shorter than some percentage of the speed of light in fiber accounting for the distance and processing time. On the other hand, the transmission time cannot be infinite, as it does not make practical sense. The variations in the transmission time are the function of transmission time so they are bound by the same rules as the transmission delay. The packet loss may range from 0 percent to 100 percent, and availability could be 0 percent, 100 percent, or some number in between. High CoS classes of service have these metrics as close to the best available value as possible (shortest latency, smallest latency variation, 0 percent loss, or 100 percent availability). Lower CoS classes have these metrics somewhat compromised.

Metrics (with the exception of availability and frame delay variation in any of its forms) are measured for a single packet [66]. In real-life networks, one gets a series of single packet measurements or a population of measurements termed a population of interest. Thus, the question arises what number we should associate with the specific network. Or, to say it differently, what number out of the population of measurements correctly represents the network performance—of course, "correctly" may mean many things here. One has a choice of maximum, minimum, average, or some percentage of collected measurements. The selection of the representative statistics may depend on the objectives of measurements—whether what we want to characterize is the behavior of the network, service, or short- or long-term traffic properties. The choices are critical for the service definition. The problem of the selection of the proper statistics representing the network status is illustrated in Figure 3.48. The upper graph provides the individual delay measurements between the UNIs over certain time period. The lower graph represents the same measurements in the form of a histogram. The type of distribution of individual measurements statistics such as average or maximum value, may be misleading if used to characterize the network behavior. The specific percentage may work best but will still miss the critical information about the maximum experienced delay.

The maximum value of a metric by itself will not reflect the network performance accurately. On the other hand, the percentage will miss the important aspect of the network behavior, which is the maximum delay.

Metrics have been defined in many standard documents. And with the variety of definition of network performance metrics, it is important to specify

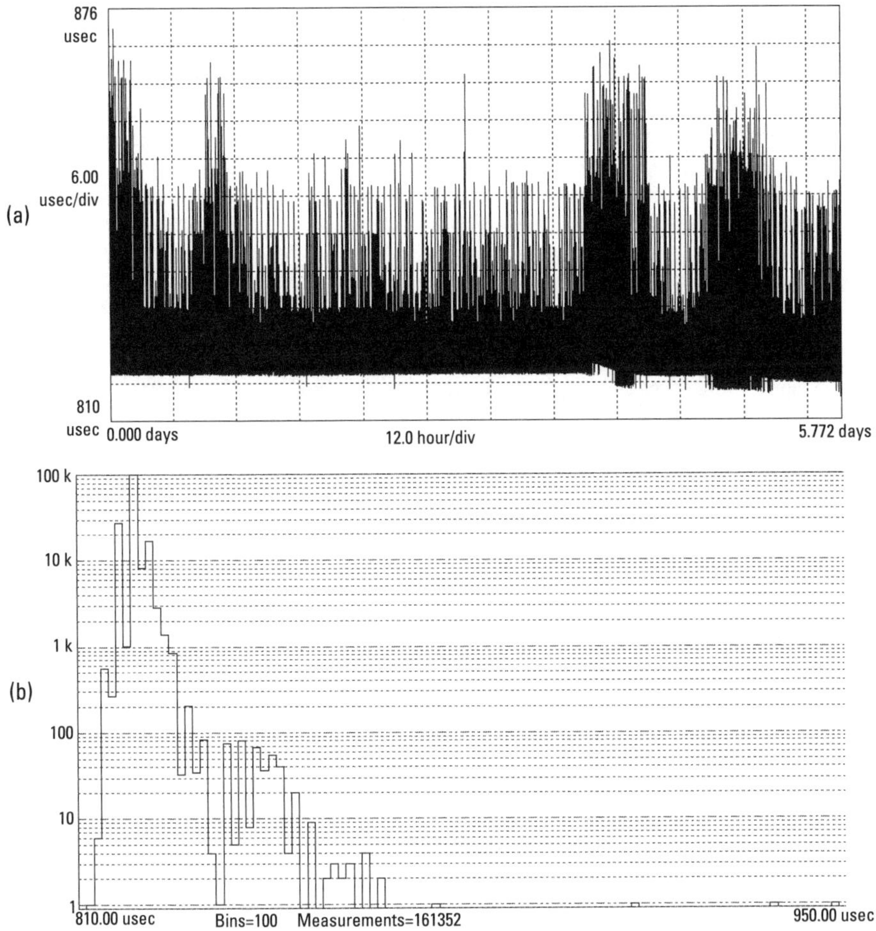

Figure 3.48 Simulated histogram of delay (FTD) measurements over time period [67].

clearly what definition one uses for the specific service. For reference, Table 3.26 provides a summary of metrics and different names to which they are commonly referred.

Of course, the standard-defined metrics have well-defined meaning. Yet, in the common parlance, it is quite often to overlook these nuances in definitions and use seemingly synonymous terms as equivalent.

The following definitions are based on the ITU-T Y.1563 standard. We define four network performance metrics: frame transfer delay (FTD), frame delay variation (FDV), frame loss ratio (FLR), and availability (Av). The definitions provided here have more descriptive meanings and are less formal than the ones in the ITU-T Y.1563 standard. This simplification is done to make them

Table 3.26
Network Performance Metrics

Network Characteristics	ITU Metric	Commonly Used Synonymous Terms [68]	Definition	Measures of Populations of Interest
Transmission time	Frame transfer delay (FTD)	Latency, delay	Transfer delay in network	Maximum, minimum, average, percentile over the population of interest
Lost frames	Frame loss ratio (FLR)	Could be defined as ratio of lost or delivered frames or packets	Lost transmission units—packets, frames	Percent of lost packets over a period of time *T*
Variation in transmission time	Frame delay variation (FDV)	Jitter, delay variation, frame delay range	Changes in transmission time	Maximum, minimum, average, percentile over the population of interest
Availability	Availability	Availability	Periods of network usability	Availability over a period of time *T*

more comprehensible to the reader. However, rest assured, nothing from the essence of the metric is lost.

3.5.1.1 Frame Transfer Delay (FTD)

Frame transfer delay is a measure of the frame transfer time between the ingress and egress UNIs (see ITU Y-1563 for a detailed discussion of the concept of delay). The reference model for the frame transfer delay metric in Ethernet networks is provided in Figure 3.49.

There are the source (SRC) and destination (DST) bridges, and bridges that are at the edge of the network. The measurement points (MPs) that define the segment of the network over which the metric is applicable are located at the edges of the bridge devices. The delay is defined between two points FRE1 (SRC) and FRE2 (DST) measurement points. Its full definition is given here:

Ethernet frame transfer delay is defined for all successful and errored frames outcomes across a Transport Operator Network. FTD is the time, $(t_2 - t_1)$ *between the occurrence of two corresponding Ethernet frame reference events, ingress event* FRE_1 *at time* t_1 *and egress event* FRE_2 *at time* t_2, *where* $(t_2 > t_1)$ *and* $(t_2 - t_1) \leq T_{max}$. *The end-to-end Ethernet frame transfer delay is the one-way delay between the MP at the SRC and DST.*

The definition excludes the errored outcomes that account for errored, lost, and spurious frames; in general, the definition does not account for frames that are rejected by the source for some technical reason. T_{max}. parameter excludes frames arriving after a certain time limit (i.e., late frames).

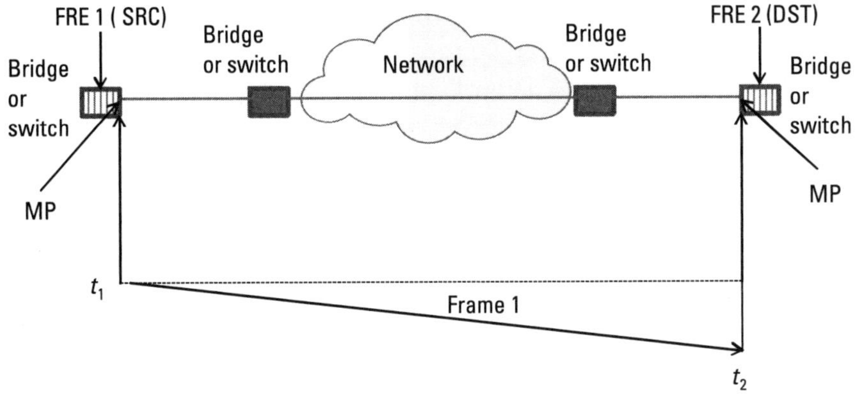

Figure 3.49 Frame transfer in Ethernet network.

Based on this definition, several metrics capturing transfer delay in the network may be defined. The transfer delay may be measured from SRC to DST or from SRC to DST to SRC. Such a transfer delay from one point to another is called one-way transfer delay. The transfer delay measured for the SRC MP packet only (on the path from SRC and DST and from DST to SRC) is called a round-trip delay. This metric is sometimes referred to as two-way transfer delay. The FDT is reported usually in a fraction of second (millisecond, microsecond, and so on).

3.5.1.2 Frame Loss Ratio (FLR)

Frame loss ratio metrics measure the ratio of the number of frames sent to the specific SRC to the specific DST point to the number of frames received at the DST minus the number of frames sent at this destination point from this SRC. The ITU Y.1563 defines FLR as follows:

Frame Loss Ratio is the ratio of total lost Ethernet frame outcomes to total transmitted Ethernet frames in a population of interest.

As with FTD, we may define one-way or two-way FLR. We can use different time intervals (i.e., populations) for the calculation of FLR like one hour, twenty-four hours, or one month. We can also use populations defined as one EVC, multiple EVCs, a port, or a number of ports. The FLR is usually calculated only on successfully delivered frames, excluding errored ones or other types of frames associated with the errored outcomes, as was the case with the FTD metric. The separate metrics for the ratio of different types of errored frames are defined [69].

Quite often instead of FLR the metric called frame delivery ratio (FDR) is reported. This is a percentage of successfully delivered frames and can be calculated by the following formula:

$$\text{FDR} = 100 - \text{FLR}$$

The FDR is report as a decimal fraction of 1 or as percentage.

3.5.1.3 Frame Delay Variation (FDV)

The frame delay variation measures differences between FTD measures for specific frames or between the TFD of a frame and some reference value. The type of the FDV metric depends on how we choose frames to compare. If the delay variation is calculated between two frames, it is called a two-point frame delay. The ITU defines the two-point FDV as follows:

The 2-point frame delay variation (v_k) for an Ethernet frame k between SRC and DST is the difference between the absolute Ethernet frame transfer delay (x_k) of frame k and a defined reference Ethernet frame transfer delay, $d_{1,2}$, between those same $v_k = x_k - d_{1,2}$.

In the case of two consecutive frames, the reference frame is the next frame arriving. This is illustrated in Figure 3.50. Frames i, j, and k have respective transfer delays x_i, x_j, and x_k. If we select some delay to be d—a reference transfer delay—the FDV for these packets will be $d_i = x_j - d$, $d_j = x_j - d$, and $d_k = x_k - d$.

The reference Ethernet frame transfer delay, $d_{j,i}$ is selected as a part of the definition, and it may be the same for all frames or it may be the frame transfer delay of the next frame.

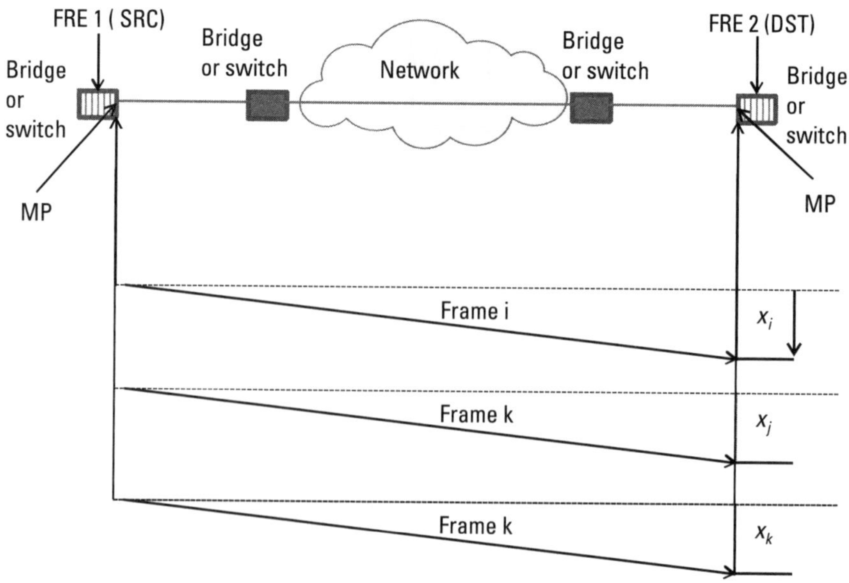

Figure 3.50 Two-point FDV.

As with other metrics, FDV may be defined over the population of interest in terms of some population statistics like minimum, maximum, average, or quantile and over different time spans (one hour, twenty-four hours, or one month) or network elements (EVC, port, pr a group of EVCs).

3.5.1.4 Availability (Av)

The availability metric is a derivative metric in the sense that it uses other metrics as part of its definition, along with a set of arbitrary parameters. Thus, it is not measuring specific physical phenomena such as latency, differences in latency, or packet losses but interprets the network behavior through these metrics.

The availability metrics attempt to capture the periods of service in which the impairments in the service are so severe that they render the service unusable or unavailable—we call it severely errored seconds (SES). Usually, these impairments are measured using the frame loss ratio metric and are evaluated over some defined time interval. The availability metric has several parameters that define periods of availability and unavailability. The parameters include the basic period of measurements (in the case in ITU Y.1563, it is one second) and the level of FLR over the basic period above which the unavailability of the basis period is declared (it is 50 percent FLR). The unavailability of the service is declared if several consecutive (10 in the ITU Y.1563 case) basic periods of unavailability are detected (10 seconds in the ITU Y.1563 metric). The unavailability period is terminated or ended retroactively at the beginning of the 10 consecutive basic periods (10 seconds) declared as available. The elements of this definition are provided in Figure 3.51.

The period of 10 consecutive SES will trigger the unavailability period. The unavailability period lasts until there is a period of 10 consecutive non-SES. Thus, the periods of availability and unavailability can be declared only *a posteriori.*

The following three sections address special topics of interest. The first section discusses the method of quick estimation of latency that can be used to evaluate the expected latency in the service. The following section discusses the service performance metrics defined by MEF. These are a bit different from the

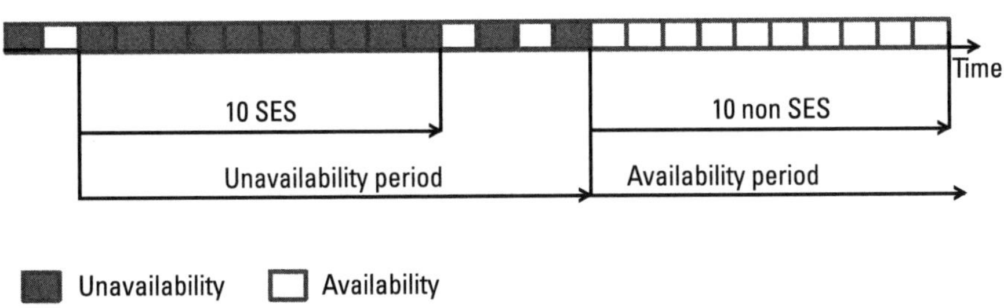

Figure 3.51 Components of availability metric (Y.1563).

ones used by ITU, and these differences need to be recognized. Finally, the last section discussed metrics for multipoint-to-multipoint services. These metrics are not so widely used for the evaluation of service and network performance. Reading this section may help in understanding why.

3.5.2 Estimating Latency

It is quite often informative in the service design to estimate expected latency (delay). Due to the nature of latency and networks, such estimates may only be cursory and will be only approximating the actual experience on the network. However, at the planning stage and possibly during troubleshooting these estimates may help in setting up the target thresholds or pointing to the excess delays in the service.

The overall delay is a sum of contributing delays from every element and segment in the path traversed by the frame. The following formula expresses this relationship:

$$D = \sum D_p + \sum D_l + \sum D_S + \sum D_q + \sum D_s$$

where:

D_p = delay due to the length of the traveled path.

D_l = delay due of the link insertion.

D_s = delay due to switching or routing.

D_q = delay due to queuing.

D_s = delay due to the stack processing.

The delay due to the traveled path is usually the largest percentage of the total delay. For fiber-based transport, it can be estimated assuming that the speed of light in fiber is about 0.68 of the speed of light in a vacuum. For practical purposes, it is assumed that 100 miles of the fiber transport imposes 1 ms latency. In estimating the actual latency in the transport, the fiber-based delay is multiplied by a factor of 1.35 or 1.4 to account for the actual fiber paths (that are never straight).

The link insertion delay is a delay incurred during the serialization of the packet at the interface. The insertion delay is the time required to send the packet from its first to its last over the interface and is dependent on the interface speed. For example, for the 1500-byte packet on a 10-Mbps interface, the insertion delay is 0.012 sec. On a 1-Gbps interface, this latency is 0.000012 sec. From the perspective of the end-to-end latency, the link insertion delay on high-speed links (1 Gbps and up) is negligible [70]. One needs to qualify this statement—in most services, the link inversion delay on interfaces greater than 1 Gbps is negligible. However, for certain high-priced services where every microsecond of delay makes a difference, the network design requires the minimization of the link insertion delay by using interfaces not otherwise justified by the require bandwidth (100G, 40G, or 10G, rather than 1G).

The switching delay is a delay incurred during the frame traversing the switch fabric. In modern equipment, this delay is in the range from 5 to 30–40 microseconds. In calculating the end-to-end path delay one needs to account for the number of hops or switches the frame must traverse. But in the metro Ethernet networks, this number is usually limited to less than 8 or 10, depending on the architecture, not adding much to the overall path delay.

Queuing delay is a delay incurred by the frame by waiting for the queues in switches to be sent to the next destination. This delay depends on the class of service and the queuing mechanism. If the frame is treated as a high-priority frame (class H), then it is processed in the so-called high priority queue and its queuing latency is minimal. If the frame is processed in queues with a limited number of processing cycles and a lower priority in the queuing schema, then a frame may include several milliseconds of delay. Queuing delay will be heavily affected by traffic congestion; in particular, this will be visible for traffic in lower traffic classes. The end-to-end queuing delay must account for all the queues in the path.

Stack processing delay is a delay due to the processing of the frame in the protocol stack. For Ethernet services, this delay is usually discounted. However, for higher-level protocols, this delay may be of significant importance to the overall delay budget. The stack delay is also of significance if the transmission media must be changed. For example, the change from wireless to wireline transport media will include a significant processing delay and will impact the end-to-end delay of the packet.

In the MEBH service, the delay in the MEBH segment is only a portion of the total delay that the packet traveling between the service end points is experiencing. The end-to-end delay and user-to-user equipment delay may be in the range 100–150 milliseconds. Only a small portion of this budget is allocated to the MEBH segment.

The composition of the performance budgets for segmented networks is another topic. One interested in it should consult the standard documents such as ITU –Y.1563.

All in all in MEN networks the largest percent of the overall delay budget can be attributed to the transport. In MEN excessive delay in the range of tens of milliseconds, if other performance metrics (jitter, packet loss) are within the target range and there is no traffic congestion, is usually caused by the excessive transport delay. In such situations one needs to verify the actual transport mileage traversed by the frame.

3.5.3 MEF Performance Metrics

MEF 10.2 defines frame delay (FD), interframe delay variation (IFDV), frame loss ratio (FLR), and availability (Av). All metrics are applicable only to the qualified service frames.

The frame delay or one-way frame delay is defined as follows:

The One-way Frame Delay for an egress Service Frame at a given UNI in the EVC is defined as the time elapsed from the reception at the ingress UNI of the first bit of the corresponding ingress Service Frame until the transmission of the last bit of the Service Frame at the given UNI [71].

The interframe delay variation (IFDV) is the difference between the one-way delays of a pair of selected service frames. This definition is in agreement with the IP packet delay variation definition given in [6] where the delay variation is defined as the difference between the one-way delay of two packets selected according to some selection function and are within a given interval [T_1, T_2].

The frame loss ratio is a ratio of packets sent to packets received minus packets sent on the UNI. This metric is defined as the one-way metric.

The definition of availability metric is similar to the definition of ITU and attempts to capture the same phenomena. Table 3.27 compares the ITU and MEF metrics.

To be accurate we need to say that MEF 10.2 defines performance metrics as "P-percentile" statistics over the population of interest of metrics (or measures) defined in the ITU sense. These are frame delay performance, frame delay range, mean frame delay, interframe delay variation performance, frame loss ratio performance, availability performance, and resiliency performance. Thus, MEF metrics differ in concept from the ITU-defined metrics. The details of metrics are provided in the MEF 10.2 document for details.

Table 3.27
ITU and MEF Network Performance Metrics

Measured Phenomena	ITU Metric Defined in ITU-T Y.1563	MEF Metric Defined in MEF 10.2	Comments
Packet transfer time	Frame transfer delay	One-way frame delay	The same metrics
Packet transfer time variation	Frame delay variation	Interframe delay variation	The MEF metric is restricted to two consecutive frames
Frame loss	Frame loss ratio	One-way frame loss ratio	The same metrics
Availability	Availability	Availability	Measuring the same phenomena but with some differences in the algorithm

3.5.4 Multipoint Metrics

The multipoint metrics measure the performance of p2mp or mp2mp services such as E-tree or E-LAN. The multipoint metrics are described in MEF 10.2 as well as in ITU–T Y.1563 documents. The specifications of metrics in these documents are not equivalent. The following is a brief summary of the multipoint metrics concepts as defined in ITU-T Y.1563. Figure 3.52 provides the reference model for the subsequent discussion.

In general it is impossible to give the precise measurements of the performance of the multipoint services, the way we can do it for the p2p service. The performance of mp2mp [72] services is presented by using the set of point-to-point measurements between UNIs of the mp2mp service and organized in matrices. The set of measurements between two UNIs belonging to the same multipoint EVC (and possible in the same multicast group) is represented as a vector of measurements for this UNI-to-UNI path. The set of measurements from one UNI to all UNIs in the multipoint EVC is a vector as well, termed the group vector. The collection of all the measurements between one UNI and all other UNIs in this EVC is the global matrix. Any statistics can be applied to any row or a column of this matrix to summarize the individual measurement results. The summary statistics can be also applied to the whole matrix of measurements to derive a global measure. The point-to-multipoint matrix can be expanded to a multidimensional matrix to represent multipoint-to-multipoint measurements. Such constructs can be used for delay, packet loss, delay range, and availability.

For the reference model on Figure 3.52 the following vector represents "m" latency measurements between UNI-1 and UNI-2:

$$\{l^1, l^2, l^i, \ldots, l^m\}_{12}$$

The series of measurement between UNI-1 and UNI-3 are represented by the following vector:

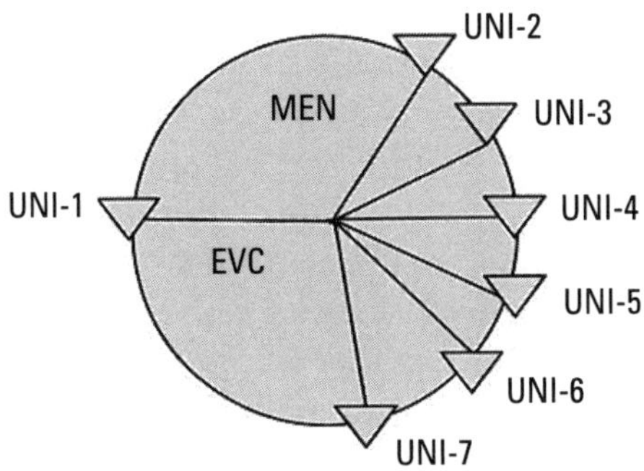

Figure 3.52 Reference model for multipoint metrics.

$$\{l^1, l^2, l^i, \ldots, l^m\}_{13}$$

This vector can be generalized to the following construct:

$$\{l^1, l^2, l^i, \ldots, l^m\}_{kp}$$

where:

i = an index of a singular data point l.

m = a number of data points in the set.

k = an index of the near end UNI in the p2p path.

p = an index of the far end UNI in the p2p path.

The complete set of data points for the EVC for UNI_1 to all other UNIs in Figure 3.52 is represented as the following matrix:

$$\left\{\begin{bmatrix} l_{12}^1, \ldots, l_{12}^m \\ l_{13}^1, \ldots, l_{13}^m \\ \ldots \\ l_{17}^1, \ldots, l_{17}^m \end{bmatrix}\right\}$$

This matrix presents the one-way measures or two-way measures, depending on the convention used. The similar constructs may be defined for traffic originating from other UNIs.

The summary statistics for every UNI I to UNI J sequence of measurements, such as average, maximum, minimum, and range of the specific metric can also be presented as a vector object or a matrix, depending on the EVC design one wants to represent.

One may define global and group statistics. The global statistics is the statistics calculated over all paths, all measurements for a given EVC, and represents, for example, the global mean delay or global packet loss. The statistic calculated for a specific index over all paths is termed the group statistics and may represent, for example, group delay or group packet loss. Provided here are only examples of summary statistics. However, any statistics can be applied to the data collected in matrices as long as they do not violate rules of algebra.

While summary positional statistics can be derived only for additive metrics, matrices of nonadditive metrics are also useful tools. For example, the matrix of availability for multipoint-to-multipoint EVC can be useful in representing the behavior of the EVC over time. Derived metrics such as the maximum or minimum of specific routes, or the percentage of nonavailable versus available segments may be very informative as well.

One needs to observe that the processing methods for the multipoint-to-multipoint metrics could be applied to provide the comprehensive summaries of the overall service composed from the several point-to-point EVCs. Such is the case if a customer having many p2p EVCs in the specific area would like to get comprehensive information about the service performance as whole.

One word of caution: there is no agreed to method to evaluate the performance of the multipoint-to-multipoint EVC. What we mean by this is that while the providers and customers can agree on the p2p metric and performance bounds for latency, packet loss, or jitter, there are no existing guidelines, for example, on how to agree to the SLAs for a group delay or global delay.

Thus, in the absence of standards, for the specific services the rules of deriving measurements and representing the results should be agreed to between service provider and the customer, using some of the provided metrics as guidance.

3.6 Service Verification

3.6.1 SOAM

One of the fundamental functions that must be supported by the networks in general and carrier Ethernet networks specifically are services related to the maintenance and diagnosis of the network and service. These functions are called operation, administration, and maintenance (OAM) services. In Ethernet networks, the OAM services at the user plane (SOAM) are defined by two standards, ITU-T Y.1731 and IEEE Std 802.1ag [73]. Going forward, we focus on ITU-T Y.1731 only. ITU-T Y.1731 expands IEEE Std 802.1ag for PM functions among others. Conceptually, however, these two standards propose a very similar architecture. Figure 3.53 presents the basic elements of the Y.1731 SOAM architecture: ME, MEG, MEP, and MIPs.

Maintenance entity (ME) represents the logical segment of the service or a network that requires management and defines the relationship between two maintenance entity group points (MEPs). Typical MEs are subscriber MEs, inter- or intraprovider MEs, or operator MEs, also defined as segment MEs or link MEs.

The MEG or ME Group specifies MEs that exists within the same administrative boundary, have the same MEG levels, and belong to the same EVC. The MEG end point (MEP) defines the end point of the MEG and can initiate or terminate OAM frames. In practical terms, MEP is a function that terminates the OAM flows and evaluates the results of OAM flows. A MIP or

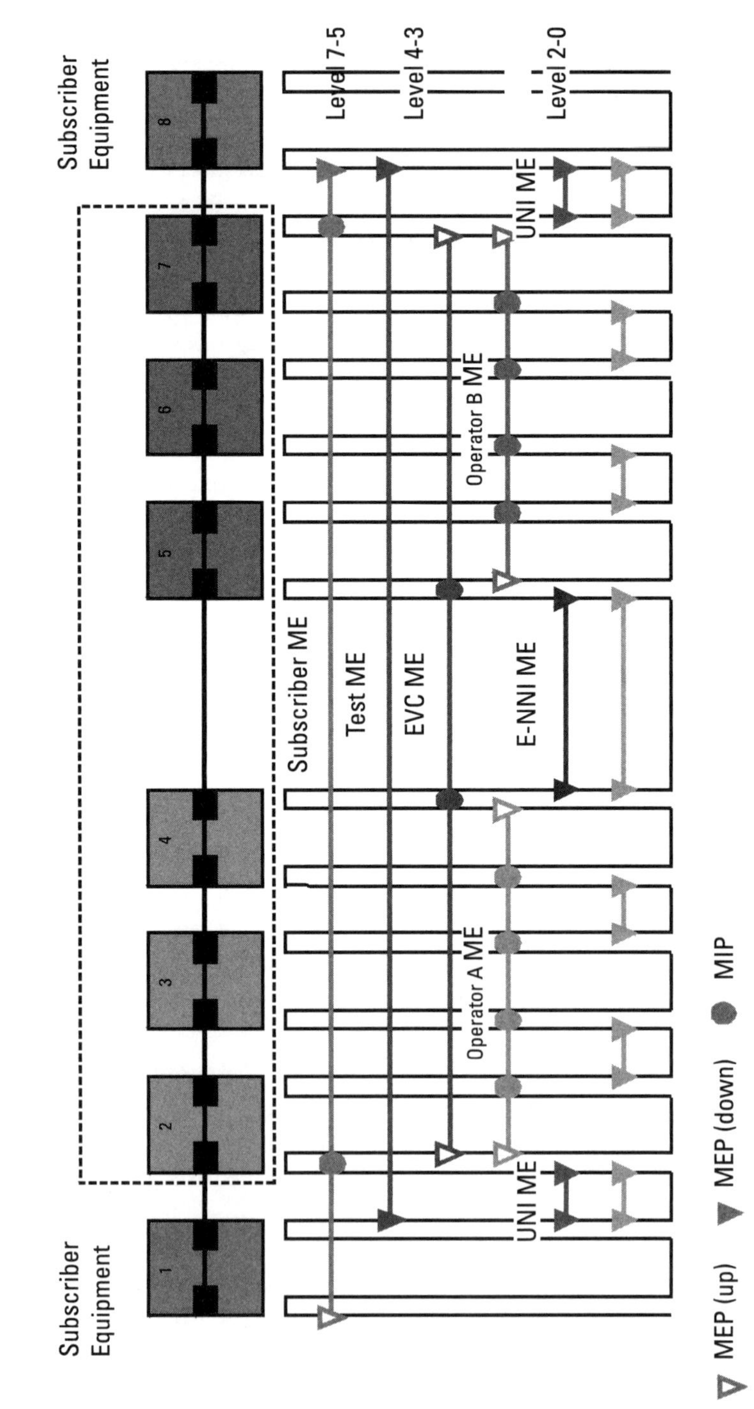

Figure 3.53 Exemplar SOAM Y.1731 architecture.

MEG intermediate point can react to some OAM frames but cannot initiate the OAM flows.

Table 3.28 presents the main SOAM functions supported in each of the standards. SOAM services are grouped into two classes: connectivity fault management (CFM) and performance monitoring (PM).

Note: The list of OAM services does not include such services as automated protection switching (APS) or maintenance communication channel (MCC), among others.

In Table 3.28 the Method column defines the specific message type used by the Y.1731 protocol to support the SOAM function.

The fault detection function is used to detect loss of continuity between the MEP points. It is implemented using the stream of CCM-type OAM frames sent at the specific intervals.

The fault verification/loopback function is used on demand to check the connectivity of the MEP to its peer or to a MIP. The connectivity is verified by sending LBM frames and monitoring the responses sent by the targeted MEP or MIP as LBR frames.

Fault isolation or link trace function is used on demand to detect the approximate location where the network failure occurred by providing the list of MIPs that can be accessed from the specific MEP. This function is executed using the LTM/LTR OAM frames.

Discovery function is supported by the same type of OAM frames as fault isolation function and is used to trace the path of the service.

Fault notification uses alarm indication signal (AIS) and remote defect indication (RDI) type frames. AIS frames are sent upon the detection of the continuity loss condition until the condition is cleared. RDI frames are sent by MEP to its peers when the fault condition detected by the continuity check.

Table 3.28
SOAM Functions

SOAM Function	Type	802.1ag	Y.1731	Method
Fault detection	CFM	Y	Y	CCM
Fault verification/loopback		Y	Y	LBM/LBR "ping"
Fault isolation (link trace)		Y	Y	LTM/LTR
Discovery		Y	Y	LTM/LTR and LTR multicast LBM
Fault notification		—	Y	AIS/RDI
Frame loss	PM	—	Y	CCM, LTM/LTR
Frame delay		—	Y	DM, DMM/DMR
Delay variation		—	Y	DM, DMM/DMR

The PM OAM functions such as frame loss, frame delay, and delay variation use modified OAM frames to obtain the performance metrics of the network or service.

In very simplified terms, SOAM services work in the following way. The user or an operator establishes the end points of the SOAM MEG—MEPs. The MEP points send specific SOAM frames between them and monitor their status. Depending on the status of the network or services, MEP points may initiate additional actions like alarms or event notification. The specific MEs are assigned specific levels. The levels can be nested and SOAM frames with the higher level are transparent to MEPs defining MEG on the lower level. The lower-level SOAM frames are blocked by the MEPs with assigned higher ME levels. Levels 7–5 are usually reserved for the customer. Levels 5–4 are reserved for the service provider and operators. Level below 4 are usually used for subsegments of the network or a link level monitoring (level 0). Some of the more frequently used SOAM services include

- Connectivity verification allowing checking that the EVC is able to transport traffic and generating an alarm or initiating other action if the connectivity is compromised;
- Performance monitoring allowing the collection of the network metrics over the monitored service.

The architecture in Figure 3.53 represents the network providing the service between two customer UNIs and spanning two operators (operator A and operator B). The subscriber supports two MEs on levels 7–5 that spans the network from UNI to UNI and should be transparent to the service providers. The subscriber ME MEP points are located at the CPE UNIs; one MEP is up and one is down. In addition, this ME has two MIPs on the UNI-Ns. These MIPs allows for the verification of the connectivity with the service provider. The test ME has two down MEPs at UNI-C and spans the whole EVC from the customer CPEs. The service provider supports the ME (EVC ME) at levels 4–3 spanning the provider network. The EVC ME is demarked by two up MEPs. It also has two MIPs at the interfaces facing the link between two operators. Each operator has its own ME with up MEPs located at the UNI and ENNI and MIPS at every switch in the route. There is also link-level MEs at UNIs ENNIs. These are down MEPs at levels 2–0.

3.6.2 Service Level Assurance

Service level assurance (SLA) specifications define service performance targets that the service is designed to support. In other words, SLAs are network performance metrics with commitment. It is important to understand the difference between the SLA metrics and the network performance metrics. The network performance metrics have an objective to characterize the behavior of the network as objectively as possible. The SLAs are service guarantees that the carrier of the service agreed to live up to as a part of the service. Violations of SLAs are usually associated with some penalties. However, you cannot violate network performance metrics. They are what they are. In fact to be precise we should have a metric_{NET} to provide the network performance and a metric_{SLA} associated with the SLAs.

Usually, the SLAs include the following metrics: FTD, FDV, and FDR. Sometimes availability is also included as a part of the SLA guarantees. In addition to the metric itself, each SLA metric has associated several attributes such as reported statics, reported population of interest and details of the measurements and the scope of the metric, and conditions of violations of the SLAs.

As depicted in Figure 3.54 SLAs are defined over the service scope under the control of the service provider. It is a scope of the EVC, which is UNI to UNI. In Figure 3.54 the service domain is defined as the performance monitoring (PM) segment. This segment spans the UNIs delimiting the EVC. More specifically, the PM segment extends between the demark service points. UNI is not the best determinant of the SLA scope as the UNI is not a physical object and SLAs have a strong physical expression in the network.

It is more accurate to define the scope of SLAs as being from the UNI-N [74] to UNI-N. The UNI-N is a part of the UNI function that resides on the service provider edge equipment. In practical terms, the scope of SLAs is defined as being delimited by the handoff ports of the service provider.

If the service is using Y.1731 SOAM facilities, the demark points of the SLAs are defined as the location of the MEP points at the UNI-N between which the PM monitoring is activated.

SLA metrics can be monitored either by using synthetic test frames or, for some metrics, by monitoring the service frames. The synthetic frames are injected into the flow of service frames at the measurement points and are retrieved at

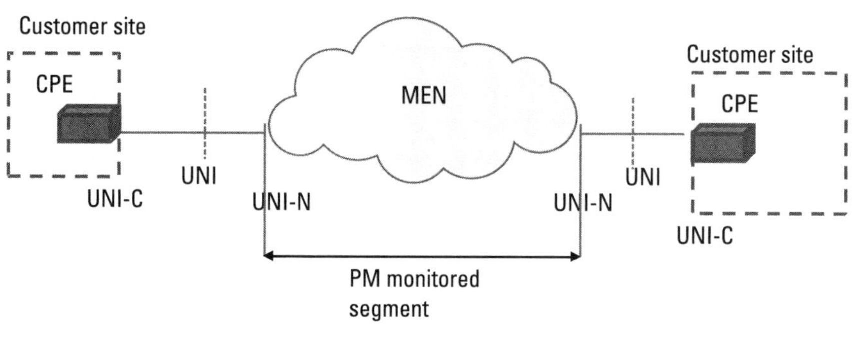

Figure 3.54 Scope of the SLAs metrics.

the handoff point of the service. There are several methods to actually accomplish it. The problem with synthetic measurements is that these are obviously representative measurements and as such have to be reinterpreted as the metrics of the actual service flows. The synthetic traffic is to be precise a sampling traffic of the actual measured flow. Some metrics such as frame delivery ratio can be measured using the actual service frames. However, such measurements are complex and rarely implemented.

Using synthetic frames poses several questions, such as what kind of frames should be used for measurements, how frequently should such measurements be performed, what is the actual metric used to represent the performance of the service, and what is the proper size of the population over which the metric is defined. All of these attributes represent slightly different aspects of the network behavior, and careful consideration should be given to each of them before the SLAs are defined or agreed to. The definition of the SLAs for the specific service should cover and explicate all of these dimensions of SLAs. Table 3.29 summarizes the exemplar structure of the SLAs.

Note: One needs to differentiate two terms—service level assurance and service level objectives (SLOs). SLAs are performance objectives agreed to by the service provider with the customer and have attached financial (or contractual) penalties. That is why SLAs are usually defined to account for a poor performance of the network with the objectives to guarantee the customer a good service but at the same time to protect the service provider from excessive penalties. SLOs are usually defined as service provider internal objectives under which the network operates. There are no penalties attached to them. They are used to gauge the performance of the network and provide the feedback on the network services status to the internal service provider engineering teams. SLOs are usually much more stringent than SLAs and rarely disclosed—they are proprietary information.

Table 3.29

Structure of SLAs [75]

Attribute	Definition
Metric	FDT, FDR, FLR, FDV, Av
Measure	Maximum, minimum, median, quantile
Population of interest	15 minutes, 1 hour, 24 hours, 1 month
SLA EVC scope	EVC, segment
SLA service scope	UNI-N–UNI-N, subsegment
Monitoring	EVC, network, service
EVC type	p2p, mp2mp
Measurement methods	Frame frequency, frame format/type, frame size

3.7 Service Interconnectivity (Engineering Requirements)

Service interconnectivity requirements are often referred to as engineering requirements. They address the physical aspects of the connectivity between the provider and the customer networks and at the design stage that the book is discussing, they are usually limited to specifications of physical interfaces, physical media, and physical aspects of the devices at the point of interconnectivity. Elements of the design included in these requirements are depicted in Figure 3.55.

3.7.1 Transport Media

At the interconnection point we usually have the choice of two types of media: copper and fiber. Cooper means the twisted pair cable. Fiber means optical fiber. MEBH requirements call for fiber only in the access due to several factors already discussed in the introduction.

With respect to fiber, one has choices between multimode or single mode fiber connections. Each choice has limitations and advantages. At the high-level design stage, the specifications are usually limited to the listing of admissible options. However, the choice of the media will impact an architecture by eliminating from the possible solution digital subscriber line (xDSL) or other copper-based access technologies or wireless links.

3.7.2 Interfaces

Tables 3.30, 3.31, and 3.32 provide the specifications of the 100 Mbps, 1000 Mbps, and 10 GiGE fiber transmission technologies. This list of available interface options is not exhaustive. These are type of interfaces currently used for the

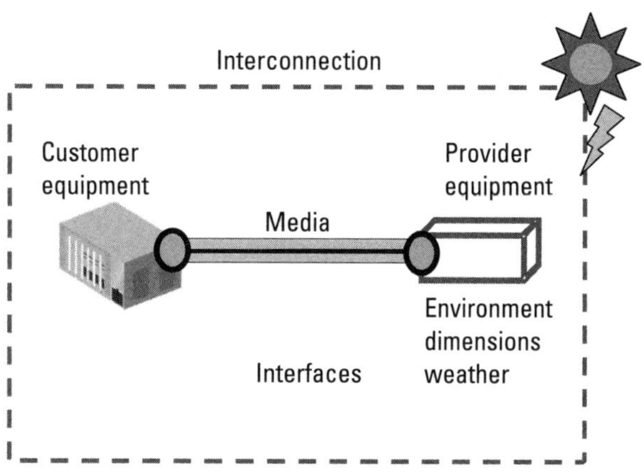

Figure 3.55 Elements of the interconnectivity design.

Table 3.30
Selected 100-Mbps Fiber Transmission Technologies

Name	Medium	Specified Distance	
100BASE-FX	Single-mode fiber (SMF) 1310 nm;	400m	IEEE Standard 802.3-2005
100BASE-SX	Multimode fiber (MMF) 850 nm	550m (1800 ft)	
100BASE-LX	SMF; 1310 nm	10 km (6.2 miles)	

Table 3.31
Selected 1000-Mbps Fiber Transmission Technologies

Name	Medium	Specified Distance	
1000BASE-SX	MMF	Up to 550m	IEEE Standard 802.3-2008 Section 3
1000BASE-LX	MMF, SMF	550m	
1000BASE-LX	SMF	5 km	
1000BASE-LX10	SMF using 1,310-nm wavelength	10 km	
1000BASE-EX	SMF at 1310-nm wavelength	~ 40 km	ITU-T G.652 SMF as specified by the IEEE Standard 802.3z .
1000BASE-ZX	SMF at 1550-nm wavelength	~ 70 km	Nonstandard

Table 3.32
Selected 10,000-Mbps Fiber Transmission Technologies

Name	Medium	Specified Distance	
10GBASE-SR	MMF 850 nm	< 400 meters, depending on cable	IEEE Std 802.3-2008
10GBASE-LR	SMF 1310 nm	10 km (6.2 kf)	
10GBASE-ER	SMF 1550 nm	30 km	
10GBASE-ZR	SMF 1550 nm	80 km	

UNI at the cells site and MTSO sites. The future will bring 40 and 100 GiGe UNIs into the service.

3.7.3 Physical Aspects

The requirements in this group define the properties of the equipment used for the interconnection such as physical dimension of the equipment, weather resistance, resistance of electromagnetic impulse, required power supply, and in-building requirements. Such requirements do not change or influence high-level architecture design, and therefore they are omitted from this discussion.

Endnotes

[1] Sir George Bidell Airy, about the potential value of the analytical engine invented by Charles Babbage. c. 1842. Quoted in Cerf, C., and V. Navansky, *The Experts Speak,* New York: Pantheon Books, 1984.

[2] MEF 10.2.

[3] Quite often point-to-point, multipoint- to-multipoint, and point-to-multipoint services are referred to as p2p, mp2mp, and p2mp, respectively.

[4] A service frame is an Ethernet frame transmitted across the UNI toward the service provider or an Ethernet frame transmitted across the UNI toward the subscriber. MEF 6.1.

[5] Hence, the concept of virtual versus physical; in the virtual separation the flows share the physical media but they preserve the logical separateness. One may say that virtual means "imitating physical."

[6] The EVC map is an association of UNIs with the specific EVC.

[7] A storm is an uncontrolled creation and transmission of BMU frames in the network due to misconfiguration. If not contained by special filters limiting the amount of BMU traffic, storms can bring the network down. In their nature BMU storms are similar to DOS attacks. The hallmark of carrier network services is the separation of the service provider control plane from the customer control plane. If such a separation does not exist, the provider's network may not qualify as the carrier network.

[8] The performance targets of services may be impacted if the services are deployed over multiple providers. In such cases, each of the providers is responsible for her specific OVC, and each OVC should be bound by specific SLAs objectives tailored to the service.

[9] This section has been developed based on the MEF 6.1.1 document.

[10] The term *discard* is defined in MEF 10.2, Section 7.13.1.

[11] The term *peer* is defined in MEF 10.2, Section 7.13.2.

[12] The term *tunneling* for L2CP frames is defined in MEF 10.2, Section 6.7.

[13] MEF 6.1.1 refers to the Metro Ethernet Forum document MEF 6.1.1, April 2012.

[14] IEEE Std 802.1D-2004, "Part 3: Media Access Control (MAC) Bridges."

[15] IEEE Std 802.1Q – 2005, "Virtual Bridged Local Area Networks."

[16] STP/RSTP/MSTP are 802.2 LLC frames, not Ethernet II type frames, and are determined by the LLC header information, not Ethertype and subtype.

[17] IEEE Std 802.3-2005, "Part 3: Carrier sense multiple access with collision detection (CSMA/CD) access method and physical layer specifications."

[18] IEEE Std 802. 1X – 2004, "Port-Based Network Access Control."

[19] MEF Technical Specification MEF 16, "Ethernet Local Management Interface," January 2006.

[20] IEEE Std 802. 1AB-2005, "Station and Media Access Control, Connectivity Discovery."

[21] IEEE Std 1588v2, TM-2008. IEEE Standard for a Precision Clock Synchronization Protocol for Networked Measurement and Control Systems, 27 March 2008, Annex F.

[22] ITU G.8264-2008. Distribution of Timing Through Packet Networks, October 10, 2008.

[23] The diagram is taken after MEF 6.1.1.

[24] Since not all CEs in an EP-tree service will see all BPDUs, undesirable behavior may ensue. Service providers should be careful to warn subscribers about attaching bridges to such a service and expecting xSTP to work properly.

[25] IEEE Std 802.3 is a collection of IEEE standards defining the physical layer and data link layer's media access control (MAC) of wired Ethernet.

[26] IEEE Std 802.1 is a collection of IEEE standards defining overall network management protocol layers above the MAC and LLC layers like VLAN protocols, PBB, PBB-TE CFM, MRP, VLAN bridging, provider bridging, LAG, MAC security, and many others.

[27] This protocol stack is a generic representation of network layers. There are other representations, more detailed such as OSI International Standards Organization (ISO) protocol stack defined in 7498-1/1994 standard. (ISO 7498-1: Information Technology—Open Systems Interconnection—Basic Reference Model: The Basic Model.) Also defined in ITU-T X.200, 07/94, Data Networks and Open System Communications, Open System Interconnection—Basic Reference Model: The Basic Model.

[28] For the complete definition of functions supported by each of the Ethernet layers, one should consult the IEEE 802.3 standards.

[29] Multiprotocol label switching (MPLS).

[30] The GFP mapping mechanism is defined by ITU-T G.7041/Y.1303, January 2002: Generic Framing Procedure.

[31] Optical carrier transmission rates are a standardized set of specifications of transmission bandwidth for digital signals that can be carried on synchronous optical networking (SONET) fiber optic networks. Transmission rates are defined by rate of the bitstream of the digital signal and are designated by hyphenation of the acronym OC and an integer value of the multiple of the basic unit of rate (e.g., OC-48). The base unit is 51.84 Mbps. Thus, the speed of optical-carrier-classified lines labeled as OC-n is $n \times 51.84$ Mbps.

[32] Refer to Section 3.4.4 for a detailed discussion of the Ethernet throughput concepts.

[33] ITU-T G.707/Y.1322, October 2000: Network Node Interface for the Synchronous Digital Hierarchy (G707); ITU-T G.783, October 2000: Characteristics of SDH Equipment Functional Blocks (G783); ITU-T G.803, March 2000: Architecture of Transport Networks Based on SDH. (G803); T G.805, March 2000: Generic Functional Architecture of Transport Networks (G805); T G.7041/Y1303, January 2002: Generic Framing Procedure (G7041); ANSI T1.105.0x SONET; ANSI T1.119.0x.

[34] DOSCiS specifications can be obtained from the Cable Labs website at http://www.cablelabs.com/cablemodem.

[35] ITU Recommendation G.709/Y.1331, Interfaces for the Optical Transport Network (OTN), March 2003 (Amendment1 December 2003); ITU Recommendation G.798,

Characteristics of Optical Transport Network Hierarchy Equipment Functional Blocks, June 2004 (Erratum 1 May 2005).

[36] DSL Evolution Technology. ADSL2/ADSL2plus/ADSL-RE/VDSL2. Broadband Forum (2008), http://www.broadband-forum.org/downloads/About_DSL.pdf.

[37] 802.3ah-2004.

[38] An amendment IEEE 802.3av to IEEE 802.3.

[39] ITU G.984.

[40] MPLS architecture is defined in RFC 3031, Multiprotocol Label Switching Architecture, January 2001, IETF. Of course, there is a multitude of RFC documents following RFC 3031 that define many aspects of the MPLS architecture. As well, there are plenty of books on MPSL technology. Here is sampling of few fairly complete ones. Evans, J., and C. Filsfils, *Deploying IP and MPLS QoS for Multiservice Networks: Theory and Practice,* San Francisco, CA: Morgan Kaufmann, 2007. Vinod, J., and S. Mulugu, *Deploying Next Generation Multicast-Enabled Applications: Label Switched Multicast for MPLS VPNs, VPLS, and Wholesale Ethernet,* San Francisco, CA: Morgan Kaufmann, 2011. Minei, I., and J. Lucek, *MPLS-Enabled Applications: Emerging Developments and New Technologies,* Chichester, UK: Wiley, 2005. Vendors publish interesting, how-to-do books about their wares. They often offer very good, general introductory sections worth studying before they open the gates to the CLI purgatory (as some would say).

[41] PBB architecture is defined in IEEE Std 802.1ah-2008.

[42] Pseudowires is a complex technology allowing for transport of layer 2 technologies over an MPLS-enabled network. The architecture of pseudowires is defined in a series IETF and ITU standards: RFC 3985 Pseudo Wire Emulation Edge-to-Edge (PWE3) Architecture. RFC 4385 Pseudowire Emulation Edge-to-Edge (PWE3) Control Word for Use over an MPLS PSN. RFC 4448 Encapsulation Methods for Transport of Ethernet over MPLS Networks.RFC 4447 Pseudowire Setup and Maintenance—Using the Label Distribution Protocol (LDP). RFC 4553 Structure-Agnostic Time Division Multiplexing (TDM) over Packet (SAToP). RFC 4623 Pseudowire Emulation Edge-to-Edge (PWE3) Fragmentation and Reassembly. RFC 4618 Encapsulation Methods for Transport of PPP/High-Level Data Link Control (HDLC) over MPLS Networks.RFC 4619 Encapsulation Methods for Transport of Frame Relay over Multiprotocol Label Switching (MPLS) Networks. RFC 4720 Pseudowire Emulation Edge-to-Edge (PWE3) Frame Check Sequence Retention. RFC 4717 Encapsulation Methods for Transport of Asynchronous Transfer Mode (ATM) over MPLS Networks. RFC 4816 Pseudowire Emulation Edge-to-Edge (PWE3) Asynchronous Transfer Mode (ATM) Transparent Cell Transport Service. RFC 4842 Synchronous Optical Network/Synchronous Digital Hierarchy (SONET/SDH) Circuit Emulation over Packet (CEP). RFC 5087 Time Division Multiplexing over IP (TDMoIP)RFC 5086 Structure-Aware Time Division Multiplexed (TDM) Circuit Emulation Service over Packet Switched Network (CESoPSN) RFC 5085 Pseudowire Virtual Circuit Connectivity Verification (VCCV): A Control Channel for Pseudowires. RFC 5287 Control Protocol Extensions for the Setup of Time-Division Multiplexing (TDM) Pseudowires in MPLS Networks. Y.1415 Ethernet pseudowires.

[43] Terminology was developed based on Benchmarking Terminology for Protection Performance, S. Poretsky, R. Papneja, and S. Vapiwala, draft-ietf-bmwg-protection-term-06.txt, March 8, 2009.

[44] The ability of a system or component to perform its required functions under stated conditions for a specified period of time. *IEEE Standard Glossary of Software Engineering Terminology,* September 28, 1990, available at, http://www.idi.ntnu.no/grupper/su/publ/ese/ieee-Se-glossary-610.12-1990.pdf.

[45] The shared risk group concepts have been adapted from ITU-T G.7715/Y.1706 (06/2002), Architecture and Requirements for Routing in the Automatically Switched Optical Network. Interdomain routing with SRG, draft-many-ccamp-srg-01.txt, as well as with the paper "Achieving Diversity in Optical Networks Using Shared Risk Groups," http://www.cs.odu.edu/~sudheer/technical/papers/journal/SRGPaper.pdf. Accessed Sept. 3, 2011.

[46] Figure represents the generic switched network architecture with edge, access, and aggregation layers. Mesh or ring architectures could be segmented in into SRGs using the same principles illustrated here.

[47] he restoration process terminology is following MEF 2.0 specifications.

[48] The service restoration times are provided as an example and do not refer to any specific technology.

[49] Availability is defined as "The degree to which a system or component is operational and accessible when required for use. Often expressed as a probability." *IEEE Standard Glossary of Software Engineering Terminology,* September 28, 1990, available at, http://www.idi.ntnu.no/grupper/su/publ/ese/ieee-Se-glossary-610.12-1990.pdf.

[50] The good treatment of the availability in networking systems is provided in Kenyon, T., *Data Networks,* Woburn, MA: Digital Press, 2002.

[51] Two standards are used in NA for the theoretical prediction of MTBF: Telecordia Electronic Reliability Prediction, US Commercial Telecommunication Standard TR-332, Issue 6/SR-332 Issue 1, and MIL-HDBK-338B, *Electronic Reliability Design Handbook,* October 1, 1998. One may also read Pecht, M. G., and F. R. Nash, "Predicting the Reliability of Electronic Equipment," *Proceedings of the IEEE,* Vol. 82, No. 7, July 1994.

[52] One may expect the MTBF of switches in the range of 200,000–300,000 hours, individual components like SFPs up to 1,000,000 hours, and smaller devices like NIDs about 100,000 hours. These are usually theoretical numbers derived from models. MTBF of deployed facilities is calculated based on the actual experience and will vary from the theoretical estimates.

[53] For discussion of network protection and availability, see Greene, W., and B. Lancaster, "Five Nines: The Myth and the Reality," *Carrier-Grade,* Vol. 3, Issue 11, 2006. See also Purser, J., "Is MTBF Data Useless?" NetworkWorld, http://www.networkworld.com/community/node/35386. Accessed April 12, 2012.

[54] One may consult the following books on the design and analysis of networks from the perspective of availability and protection. Ideas presented in the books are not necessarily vendor-bound but of more generic import. Oggerino, C., *High Availability Network Fundamentals,* Indianapolis, IN: Cisco Press, 2001. Xu, Z., *Designing and Implementing IP/MPLS-based Ethernet Layer 2 VPN Services,* Indianapolis, IN: Wiley Publishing, 2012. A detailed book about protected network design that is not-vendor specific is Vasseur, J.-P., M. Pickavet, and P. Demeester, *Network Recovery: Protection and Restoration of Optical, SONET-SDH, IP, and MPLS,* Burlington, MA: Morgan Kaufmann, 2004.

[55] Based on Poretsky, S., R. Papneja, and S. Vapiwala, IEFTOrg, http://tools.ieft.org/html/draft-ietf-bmwg-protection-term-06.txt, March 8, 2009.

[56] A very good introduction to the QoS mechanisms is offered by a book by Tim Shigeti, *End to End QoS Network Design,* Indianapolis, IN: Cisco Press, 2005 (and you do not have to be CCIE to read the book) and from ALU a book by Balakrishnan, R., *Advanced QoS for Multi-Service IP/MPLS Networks,* Hoboken, NJ: Wiley Publishing, 2008. These are the best-practices books, not going too much into theory or abstraction. Abstractions are good in giving some general views of methods and mechanisms, but they are far cry from implementation. There is a relative abundance (on any topic, by the way) of books about abstract QoS concepts; there is a scarcity of good books on the best-practices in network engineering (and in life in general). Some people even say that abstraction is about nothing, as "no thing" means not a physical thing, something that is not. Other books that may be consulted on the QoS design are, for example *QoS Quality of Service* by Paul Ferguson and Geoff Huston, Hoboken, NJ: John Wiley, 1998; *Internet QoS: Architectures and Mechanisms for Quality of Service* by Zheng Wang, Burlington, MA: Morgan Kaufman, 2001; Kenyon, T., *Data Networks,* Woburn, MA: Digital Press, 2002. As I said, the list is endless. Generalized QoS management concepts are discussed by Chao, J., and X. Guo, *Quality of Service Control in High-Speed Networks,* New York: John Wiley and Sons, 2002.

[57] Red and yellow designations of traffic have nothing to do with any color. This is just a designation of traffic satisfying or not satisfying certain parameters of bandwidth profiles. Traffic conforming to CIR/CBS parameters is called *green* and has nothing to do with the preservation of the Earth's ecosystem.

[58] The acronym RED means random early discard, and it refers to the algorithm used to manage the queue congestion. This algorithm will be explained in the following sections.

[59] See presentation by Marshall, P. Y., 1564 Ethernet Service Activation Testing Methodology, Sunrise White Paper, February 2011 ITU-T Y.1564 Appendix I, CBS and EBS Test Methodology.

[60] Token is a specific quanta in bytes by which the *token bucket* is refreshed at the specific time intervals.

[61] The value of T_f is an example, and it does not represent any specific equipment or implementation.

[62] *Service frame* is an MEF 10.2 term denoting the customer frames with the data traffic.

[63] From "Understanding Carrier Ethernet Throughput: Am I Getting the Throughput I Should Be Getting?" Metro Ethernet Forum, March 2010.

[64] TCP throughput is a function of the window size and indirectly of the latency of the transport and frame losses. For details on the TCP throughput, one may consult Constantine, B., G. Forget, R. Geib, and R. Schrage, "RFC 6349 Framework for TCP Throughput Testing," 2011.

[65] The detailed discussion of the effect of network QoS on the TCP transmission performance is provided in ITU-1 Y.1541 Network Performance Objectives for IP-Based Services, Appendix IX.

[66] The FLR for a single packet is 0 percent or 100 percent.

[67] Examples of measurements have been obtained using Symmetricom Time Monitor Analyzer.

[68] Some terms may have a slightly different meaning.

[69] Metrics for errored frames may include the ratio or a number of runt frames, errored frames, timeout frames, long frames, jubber frames, FCS errors, CRC errors, and giant frames. The list of types of errored Ethernet frames is not complete.

[70] This, however, may not be the case if the service supports jumbo frames.

[71] MEF 10.2.

[72] The p2mp service can be treated as a special case of the mp2mp service. By extension, the p2p service is a special case of the mp2mp service as well.

[73] The reference to OAM as user plane service means that the OAM services can span the whole service segment of the network—UNI to UNI—and are not limited to the directly connected transport segments (link OAM).

[74] UNI-C and UNI-N are components of the UNI. The UNI-C functions reside on the customer equipment and the UNI-N on the service provider's equipment.

[75] The values in table are provided as examples only and do not refer to any specific service.

4

MEBH Service Design

A network design that satisfies only 100 percent of requirements is a poor design.

—*The Tao of Network Design.*

4.1 MEBH Service Logic

Service logic defines the logical connectivity between the UNI at the cell site and the UNI at the MTSO. The service logic defines the architecture of the MEBH service, defines the interconnectivity between the customer and the MEBH service provider, and determines the resiliency of the service. In short, it is of fundamental importance to the design of MEBH.

Once the MEBH service is deployed, one can change most of the service parameters or attributes: one can change bandwidth allocated to the UNIs, change QoS properties, improve performance, even change UNI configuration. However, once the service is deployed, to change its service logic or the architecture amounts more or less to the redesign of the service from ground zero. Needless to say, defining the architecture or service logic for MEBH service is of primary importance.

4.1.1 MEBH Using E-Line Service

Figure 4.1 presents four options for the MEBH service architecture using E-LINE service logic:

- Option A is the EP-line (EPL) service model.
- Option B is the EVP-line (EVPL) service model.

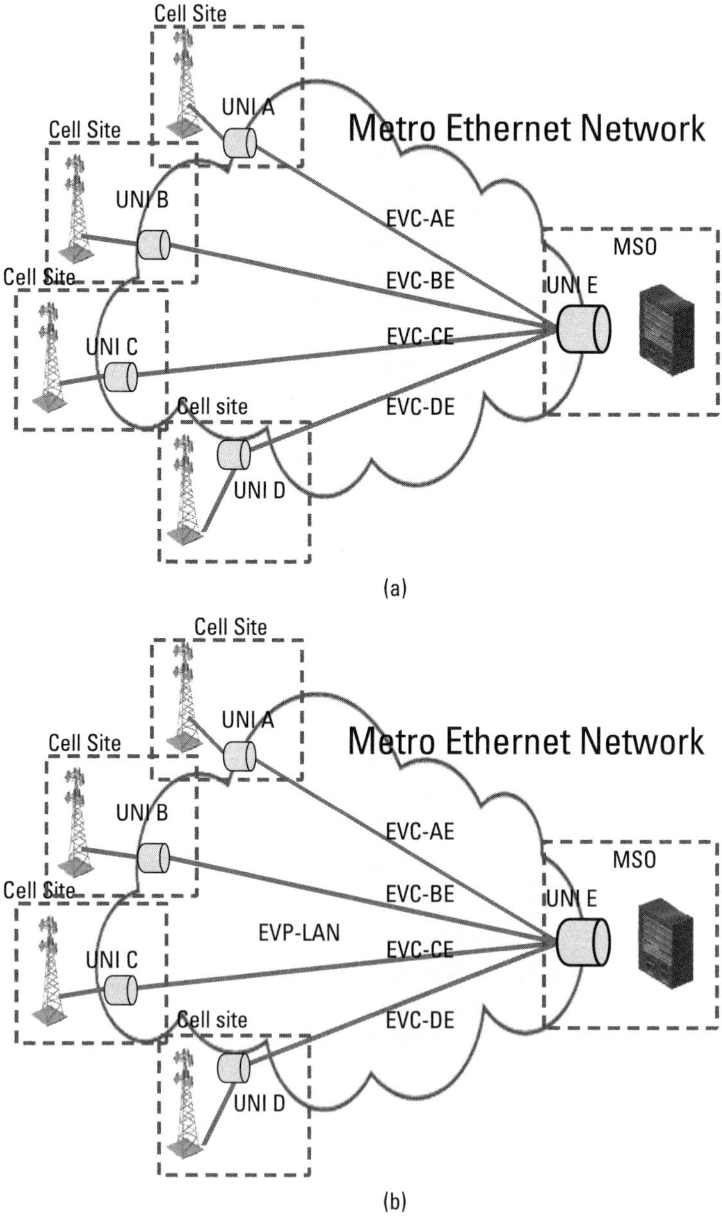

Figure 4.1 E-line-based service options for MEBH.

- Option C is the same as option B but with the protection of the MTSO handoff.
- Option D is the same as option B but with the multi-QoS service.

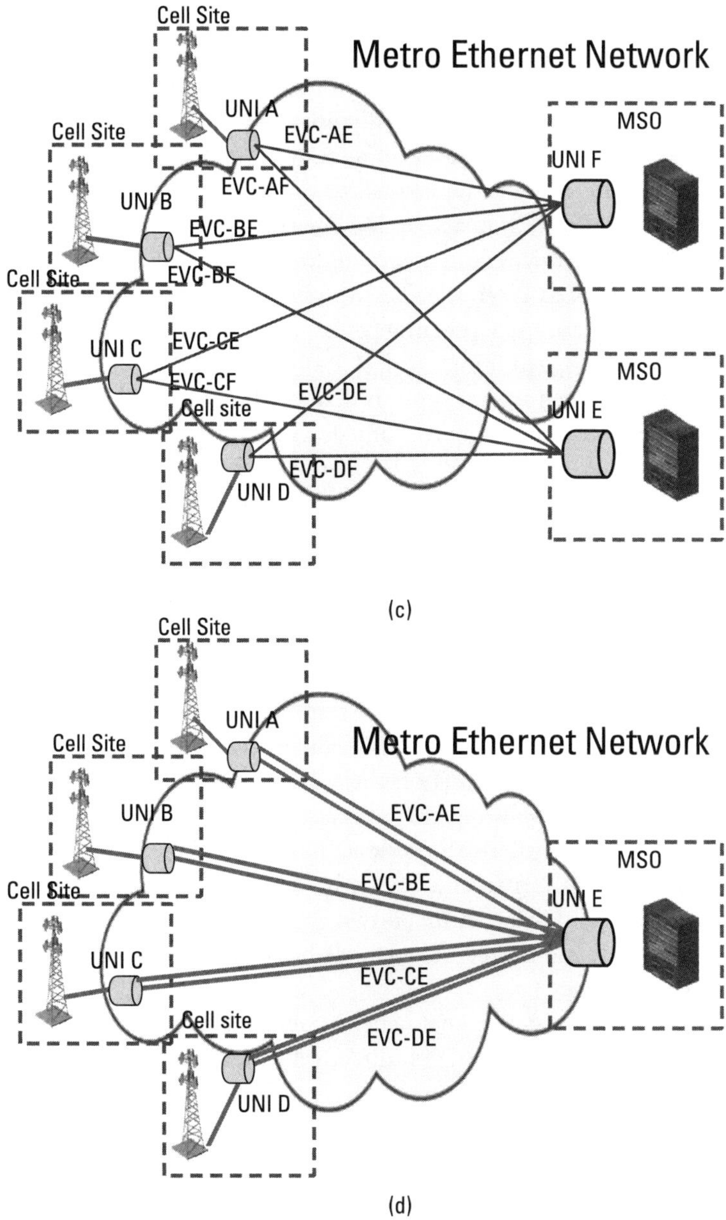

Figure 4.1 (continued)

The EP-line service (Figure 4.1(a)) is a highly transparent service supporting EVCs between two UNIs. All UNIS are all-to-one type (i.e., there is no multiplexing of traffic on any of the EPL UNIs). The service is transparent, meaning that the customer may send any type of frames, untagged or tagged, into the UNI (all-to-one UNI). However, with this service only one EVC is mapped into one UNI, which requires one physical interface for each EVC. Thus, with this service one cannot engineer the aggregated UNI with all of its benefits. This type of service is obviously not scalable and one would not see it implemented quite often. However, some providers deploy the EPL service in a way that the UNI cell site is the EPL all-to-one UNI and the MTSO UNI is a multiplexed UNI. While not strictly MEF compliant, such a service offers the benefits of the EPL service transparency and simplicity combined with the benefits of the multiplexed UNI. Needless to say, such a "mixed" service requires careful engineering and may potentially lead to configuration issues (e.g., with the treatment of untagged frames or L2CP protocols at both UNIs).

The EPL service does not support dedicated intercell site communication channels or separate management channels. All the communication to the cell site is done using the single EPL EVC between the cell site and the MTSO. There is no built-in protection or the stand-by path. The resiliency of the design depends on the resiliency of the network and network elements supporting the service. The design would not rather guarantee the resiliency above four nines.

The simplest EVPL service uses a single EVC from the cell site to the MTSO (Figure 4.1(b)). The interface at the MTSO is multiplexed—EVCs from many cells sites share this UNI. In this design, the cell site can only send and receive traffic from the MTSO UNI. The number of EVCs multiplexed on the MTSO-facing UNI depends on the UNI capacity and ability to support EVC configurations such as ingress policers. The UNI at the cell site is tagged. The EVC has a single CoS so all traffic will be mapped and transported at this CoS with the same priority. The CoS is H or M with respective SLAs for latency, packet loss, and jitter. Availability metrics may also be included. All SLA metrics are actively monitored. The choice between the H or M class is up to the wireless operator.

There is no separate channel between cell sites and MTSO for managing the cell site equipment. There is no direct cell-to-cell connection. In a case of a failure of the EVC to the particular cell site, all communication with this cell site is terminated.

The more complex EVPL-based service architecture is presented on Figure 4.1(c). In this design, each cell site has two EVCs to two MTSO locations. One EVC is designed as active or working; the other as stand by. Only one EVC is used at one specific time to transport data traffic; this EVC is designated as a working EVC. When the working EVC fails, the traffic is switched to the stand-by EVC. Both EVCs have to be operational, as the MEBH provider is not aware which one is used at any time. The detection of the EVC condition and triggering of the switching of traffic is done by the customer. The mechanism of detecting the status of the EVC and forcing switching depends on the customer's design and should be transparent to the MEBH service provider. The number of active and stand-by EVCs is not necessarily limited to two. Any number of EVCs supported by the technology may be used in this configuration. The two-EVCs design is to some extent a canonical case for the EVPL protected design.

The two EVPL EVC solution may be design with two EVCs being of the same CoS class. The alternative solution is to design one of the EVCs as "M" class and use it only as a backup EVC. This will lower the cost of the MEBH service (if there is a price differential between the H and M services). Such a solution will require revertive action on the part of the wireless providers, so the active traffic will stay on the lower CoS class EVC only temporarily.

The MEBH service architecture for mQoS EVC is presented in Figure 4.1(d). Rather than having an EVPL EVC of the single CoS class this service uses EVC with multi QoS facilities. Such an arrangement is judged to be less expensive per bit transported than the dedicated single high QoS class EVC. Of course, this assumes that the multi-QoS EVC offers a price advantage over the single, high class EVC. The design requires that the customer edge device—the CPE—performs the classification of the traffic into separate traffic classes. For this service, one needs to maintain separate SLAs per class of traffic. The other properties of this architecture apart from the ability to support multiple CoS classes over the same EVC are the same as for the single CoS EVPL EVC architecture.

4.1.2 MEBH Using E-LAN

The EVP-LAN is a multipoint-to-multipoint EVC. With this service, several options to implement the MEBH service are possible. The simple design would use the one EVC, EVP-LAN EVC, for all the cell sites as presented in Figure 4.2(a).

The service based on variants of the EVPL-LAN has properties of the multipoint-to-multipoint EVCs. In particular, all the cell sites in the EVP-LAN EVC are in one broadcast domain. The number of cell site UNIs in one EVP-LAN EVC depends on the design but in general has to account for potential broadcast issues characteristic of the multipoint-to-multipoint designs. The failure of this EVC would bring down the service to all cell sites serviced by this EVC. The recovery times are usually longer for EVP-LAN EVCs than for EVPL EVCs. But these aspects of the design are strongly implementation dependent. As well, the SLAs for multipoint-to-multipoint EVCs are more complex and less understood as the SLAs for point-to-point services (one way to address this issue is to provide point-to-point metrics between MTSO UNI and the cell site UNIs, as in EVPL EVCs). Measuring of the EVP-LAN performance would require the construct of the performance matrix as explained in the preceding sections. To contain broadcast domain and limit the impact of EVC failures, the

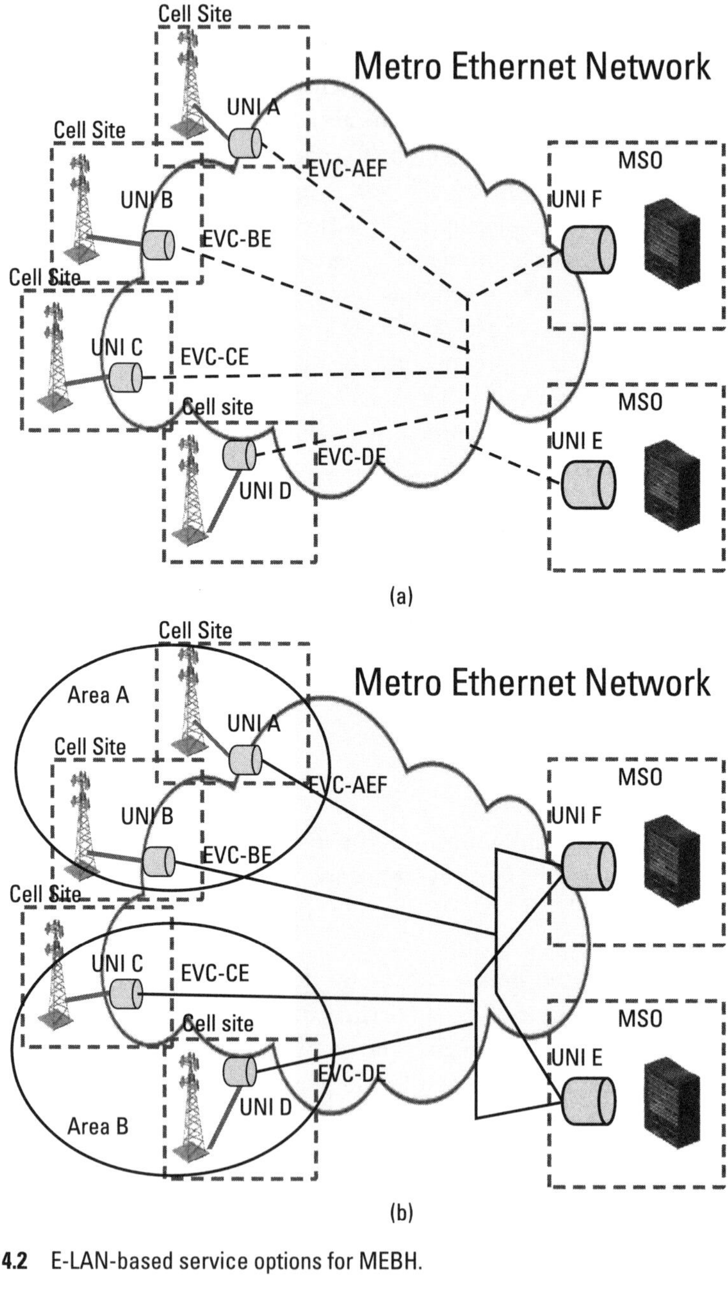

Figure 4.2 E-LAN-based service options for MEBH.

EVP-LAN designs may group in one EVC cell sites geographically collocated. This design is presented in Figure 4.2(b).

The more complex design would use one EVP-LAN EVC per cell site with two handoff UNIs to MTSO, as illustrated in Figure 4.2(c). One UNI

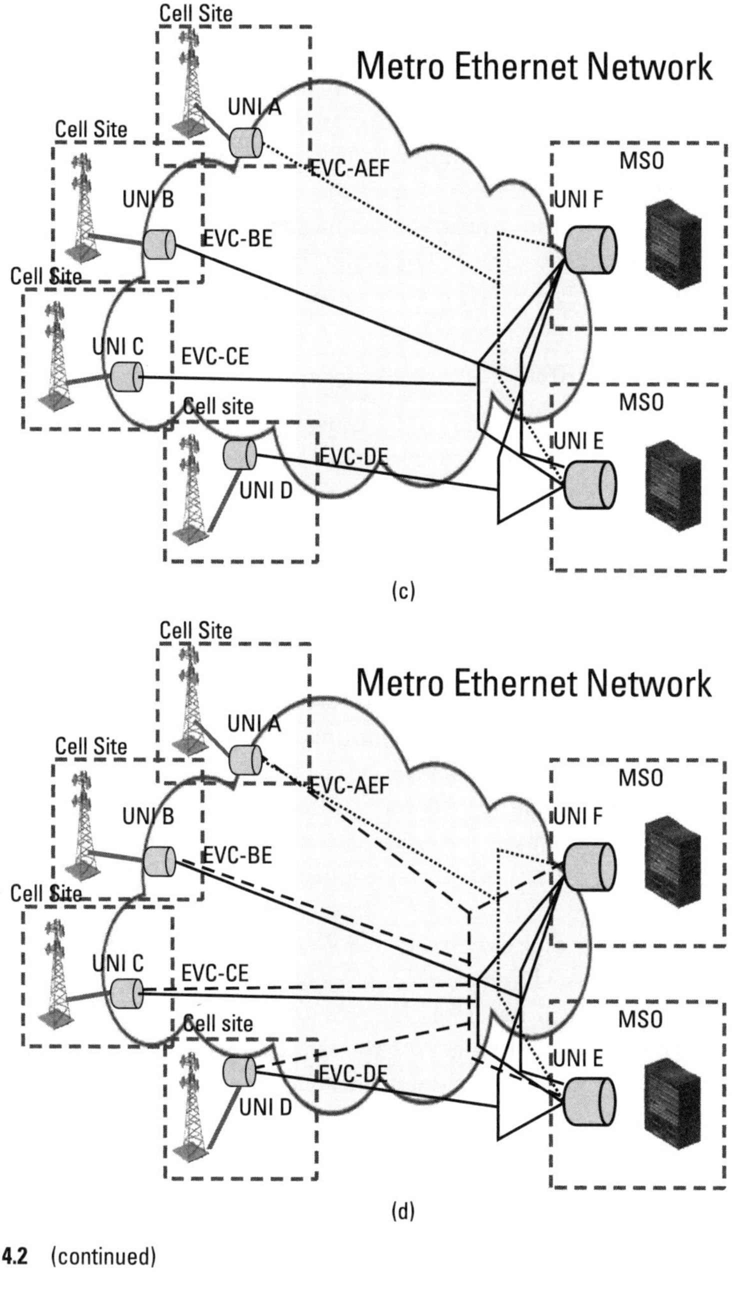

Figure 4.2 (continued)

would be at the cell site and two UNIs would be facing MTSOs. In such a design, the handoff UNI to MTSO would be protected against the failure of the handoff UNI, the handoff transport, or the CPE equipment in the MTSO. If the handoff UNI fails, the traffic is switched to the other handoff UNI. As this is a bridged service (E-LAN) traffic is switched when the new path is learned and the mapping of MAC DA to port is learned at the bridge. There is no dedicated cell site–to–cell site communication EVC. The data and management traffic is always transported from and to the cell site from the MTSO.

An alternative design would overlay on the design from Figure 4.2 (an additional EVP-LAN EVC spanning all the cell site UNIs as well and MTSO UNIs). This EVC would provide the management access to the cell sites, independent of the data channels. The design is presented in Figure 4.2(d). The EVPL-LAN spanning all UNIs is represented with a broken line.

4.1.3 MEBH Using E-Tree

The architecture for the MEBH service using EP-tree EVCs is presented in Figure 4.3(a). The UNIs facing the cell sites are leaves of the EP-tree and the UNI facing the MTSO is a root. The EP-tree EVC provides the data channel between the cell sites and the MTSOs. The EP-tree design cell sites can communicate only with the root. The root can communicate only with cell sites. Thus, there is no separate intercell communication channel. The advantage of the EP-tree over the EP-LAN is that the EP-tree limits the broadcast and simplifies traffic flows. The EP-tree is all-to-one port service. The service in its simplest version is designed with the single CoS class. However, it can accommodate multi-QoS traffic if desired. The EP-tree service in principle is very similar to the EP-LAN service. Thus, in the actual implementations, the EP-tree is often implemented with EP-LAN by applying specific traffic filters preventing cell site–to–cell site traffic. The design in Figure 4.3(b) has two roots for protection. In case of the failure of one root, the second root may take over the role of the primary root.

The design uses EVP-tree EVCs. In this design the MEBH service is provided with multiple multiroot EVP-tree EVCs. Each EVP-tree EVC services the selected cell sites UNI (possibly in a common geographic region). Each EVP-tree EVC has also two roots for protection. Such a design provides the additional protection of the root. As the service area is divided into multiple EVCs, the failure of one EVC does not affect the status of other EVCs. Of course, this is true only if these EVCs do not belong to the same risk group on the level of the switch, link, or transport facilities. The design is single CoS, but with few modifications it can support mCoS service.

4.1.4 MEBH—Mixed Solutions

Other combinations of EVP-LAN EVCs are possible. For example, an EVP-LAN EVC may be used to provide the intercell site communication channel as depicted in Figure 4.4(a). The channel is used only for intercell site control traffic. The span of this EVC may be over all UNIs or limited to the geo-

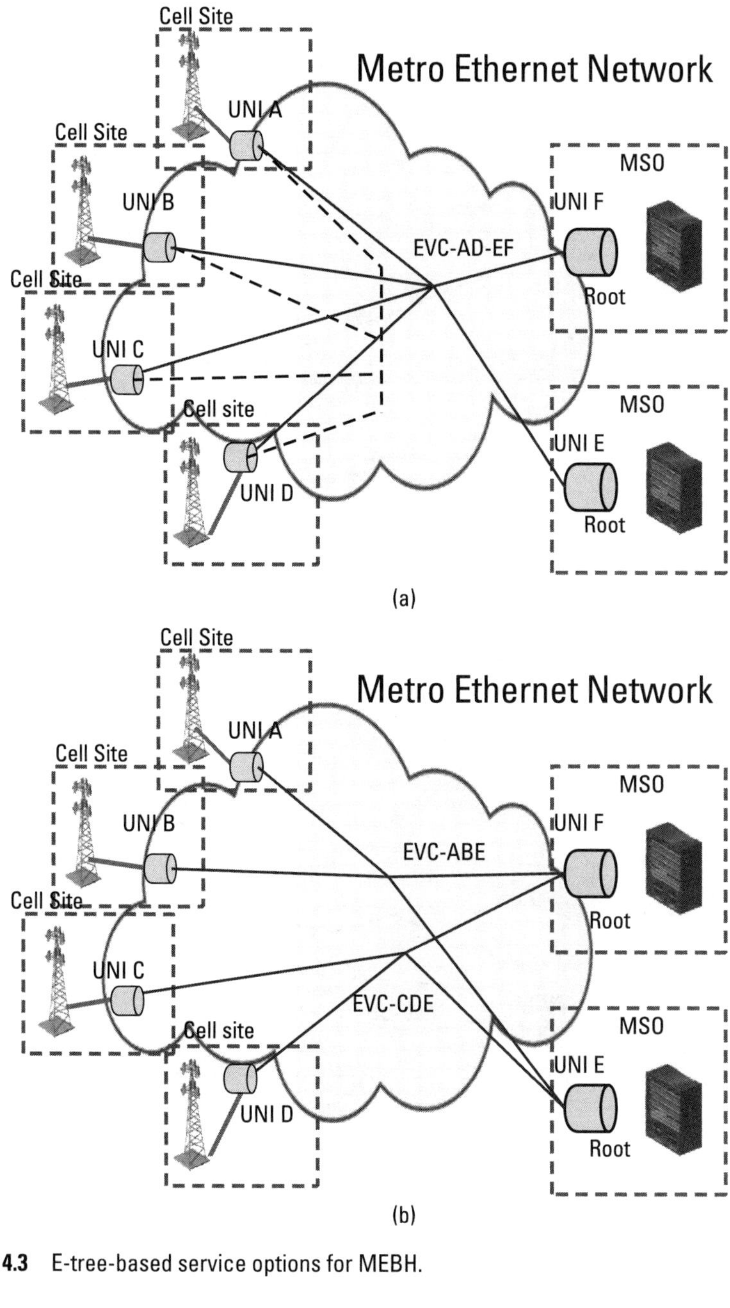

Figure 4.3 E-tree-based service options for MEBH.

graphic clusters. The MEBH service using EVP-LAN and EVPL in Figure 4.4(b) combines multipoint and point-to-point services. Each cell site has two EVPL EVCS and one EVP-LAN EVC. EVPL EVCs provide the data channel from the cell site to the MTSO. EVCs are deployed in pairs, thus providing

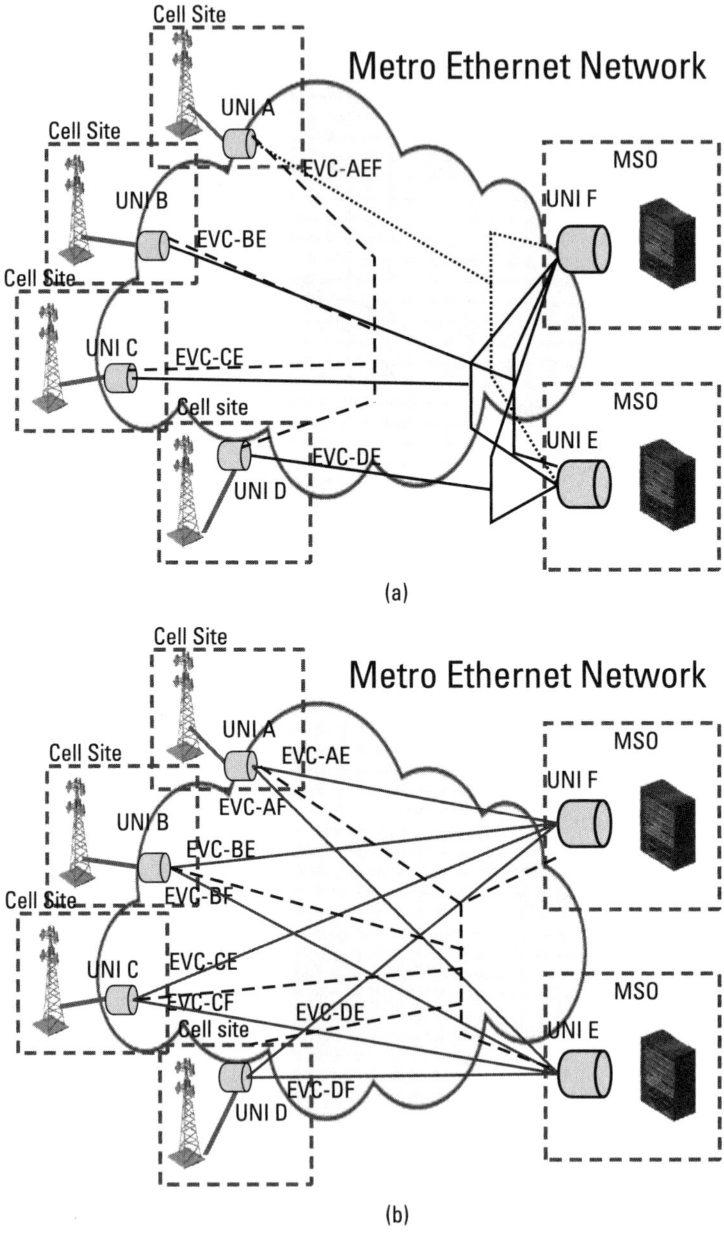

Figure 4.4 Mixed service logic MEBH designs.

the protection of the cell site and MTSO against certain types of failure of one EVC. There is no difference in how these two EVCs are provisioned; which one is active and which is passive at a given moment is dependent on the control mechanism at the CPEs in the MTSO and the cell site. The properties of the

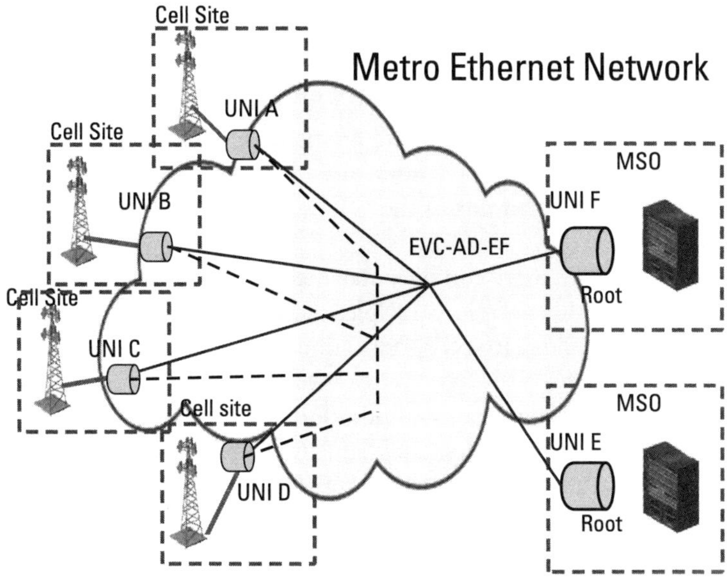

(c)

Figure 4.4 (continued)

data channels in this design are the same as for the architecture with EVPL EVCs only. However, by adding the EVP-LAN EVC acting as the management channel, this architecture adds new functionalities (an independent channel for the MTSO to cell site traffic). The EVP-LAN channel is independent of the data channel EVCs. The management traffic is separated from the data traffic. Thus, in case of the failure of the data channels, one may be able to access cell sites using this management channel. The independence of the data and management channel depend on whether they are in the same shared risk group of facilities. The management channel built using the EVP-LAN EVC allows in principle intercell site communication for a limited amount of traffic (due to the relatively small BW capacity allocated to this EVC) as this EVC is dimensioned for the management traffic only.

The third option for the mixed-service logic architecture including EVP-tree and EVPL-LAN EVCs is presented in Figure 4.4(c). This design in addition to the EVP-tree EVC with properties similar to those described in the previous section has also an EVP-LAN EVC that provides a communication channel for intercell site traffic. The UNIs in this design must be multiplexed, as they have to support two EVCs at the cell site: the EVP-LAN and the EVP-tree. The EVP-LAN may be extended and include the UNI to MTSO. This construct would provide a separate channel for the management traffic. The definitions of the EP-tree and EVP-LAN have been provided the preceding sections.

4.1.5 Multiservice Area MEBH Architecture

There can be many models of multiservice area MEBH architecture. The simplest one includes two MEN areas and EVPL service, with the MTSO located in one MEN and all of the cell sites in another. Figure 4.5 resesents this architecture. Each EVC is combined from two OVCs; each OVC is in a separate service area. The speciifcations of the OVCs and ENNI should betransparent to the wireless provider.

Designs that are more complex may include the cell sites in both MENs, EVP-LAN EVCs. However, in principle, the wireless provider should be isolated from the OVC and ENNI specifications and define only UNI and EVC attributes.

Other service architectures in addition to these presented are feasible. One may add separate management channels or cell site–to–cell site communication channels adding p2p or mp2mp EVCs. The selection of the design is dictated by the design of the MTSO services—by their ability to detect failure and switch the active traffic to working facilities.

4.1.6 L2CP Filtering

L2CP processing requirements for the proposed MEBH service architectures follow the standard requirements for the type of service as presented in Table 3.11, Table 3.12, and Table 3.13. Table 4.1 and Table 4.2 illustrate L2CP filtering rules for the EVPL service logic model. For other service logics, the filtering rules maybe created in the same way—by extracting the relevant sections from the L2CP rules tables. The difficulties come if the service logic uses different types of EVCs. As the L2CP rules are applied at the port or UNI level, for all

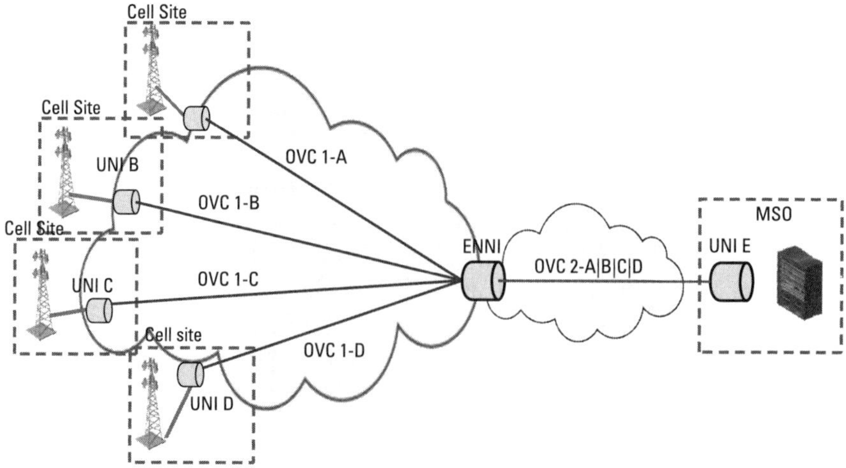

Figure 4.5 Multiservice area MEBH architecture.

Table 4.1
L2CP Processing – Step 1

Destination MAC Address	L2CP Action for EVPL, EVP-Tree, EVP-LAN
01-80-C2-00-00-00 01-80-C2-00-00-01 through 01-80-C2-00-00-0A 01-80-C2-00-00-0B 01-80-C2-00-00-0C 01-80-C2-00-00-0D 01-80-C2-00-00-0E 01-80-C2-00-00-0F	*Must not* tunnel (additional requirements may apply as per the specific service type)

Table 4.2
L2CP Processing Requirements for Virtualized Ethernet Service

Protocol Type	L2CP Action EVPL	L2CP Action EVP-LAN
STP/RSTP/MSTP	*Must* peer on all UNIs or discard on all UNIs	*Must* peer on all UNIs or discard on all UNIs
PAUSE	*Must* discard on all UNIs	*Must* discard on all UNIs
LACP/LAMP	*Must* peer or discard per UNI	*Must* peer or discard per UNI
Link OAM	*Must* peer or discard per UNI	*Must* peer or discard per UNI
Port authentication	*Must* Peer or discard per UNI	*Must* peer or discard per UNI
E-LMI	*Must* Peer or discard per UNI	*Must* peer or discard per UNI
LLDP	*Must* discard on all UNIs	*Must* discard on all UNIs
PTP peer delay	*Must* peer on all UNIs or discard on all UNIs	*Must* peer on all UNIs or discard on all UNIs
ESMC	*Must* peer or discard per UNI	*Must* peer or discard per UNI

EVCs on the UNI, for the mixed service logic designs one should select L2CP filtering rules that are most exclusive of the service logic present on the UNI (i.e., the rules that would avoid creating potentially damaging service conditions in any of the services present on the UNI).

We could restate main principles guiding the design of the service logic for MEBH services as follows:

- The selected service logic should support the protection design by offering working and stand-by channels.
- The selected service logic should facilitate the design of the distributed facilities.

- The service logic should support separation of the management and data traffic.
- The selected service logic should follow the standards, as it should result in predictable behavior.

4.2 MEBH Transport

All agreed-to requirements will be changed.
—*The Tao of Network Design.*

With service logic defined, we add to the service design transport aspects of the architecture. One may think that the customer of MEBH service has a limited choice or no choice at all in the selection of transport technologies. Alternatively, one may think that the transport technology should be a black box to customer services. Both statements are only partially true. The customer actually has several choices, and she must be made aware of consequences of each choice. As we have seen from Table 3.17, not all transport technologies are created equal. Some are more flexible than others, offering an open evolutionary path for growing services. Some technologies lock the architecture into a certain solution and logic with no path to grow. The choices that the customer has in deciding on the transport technology are as follow:

- One may build her own dedicated transport network.
- One may request from the provider that some access or core networking technologies are not to be used for the service, or one may select a provider with the technology of choice.
- One may request certain changes to the existing service transport architectures or even request changes to the design.

Each choice should be guided toward using the most advanced technologies offering the greatest flexibility and capacity to be customized. The second rule of thumb is that the selected technology should support as much as possible generic Ethernet service requirements, beyond what is required in the current design. The technology that satisfies—even perfectly—only requirements for today should not be selected; it will not allow the service to evolve.

I will avoid defining the desired MEBH transport architecture in terms of specific technologies. Such an approach would certainly cause heated religious discussions. The truth is that most of modern networking technologies can deliver solid service or be made to do so. However, they may not be optimal or future-proof. In addition, in some cases one may have to use the technology that is in the ground. Life always puts reality checks on our dreams. However, I will discuss examples of a few design options for MEBH services, and the readers can make their own choices to fit their needs. Figure 4.6 illustrates the major requirements for access, transport, and handoff segments of MEBH service. These may serve as a guide for the selection of transport technologies.

Following are a few examples of the transport design implications toward the service. Figure 4.7 presents the SONET/SDH-only MEBH solution. This design offers the dedicated bandwidth; high quality of service; protection on access, core, and handoff; and proven the SONET/SDH technology reliability. With use of Ethernet handoffs at both UNIs, not a standard offer in SONET/SDH service, the customer obtains almost a layer 2-like service. However, building into this design the layer 2 facilities like VLAN awareness, QoS, and overprovisioning requires additional service development. The service logic offered by such a solution is EPL only, with all-to-one bundled UNIs. The advantage of such service is its transparency to all layer 2 features. Shortcomings of the SONET/SDH design will be seen in limited ability to support X2 interfaces, coarse granularity of bandwidth upgrades (usually in steps of STS1-nv), no overprovisioning, and some limitations of the bandwidth due to SONET/SDH GFP overhead. Achieving the full support for CIR/CBS-defined traffic

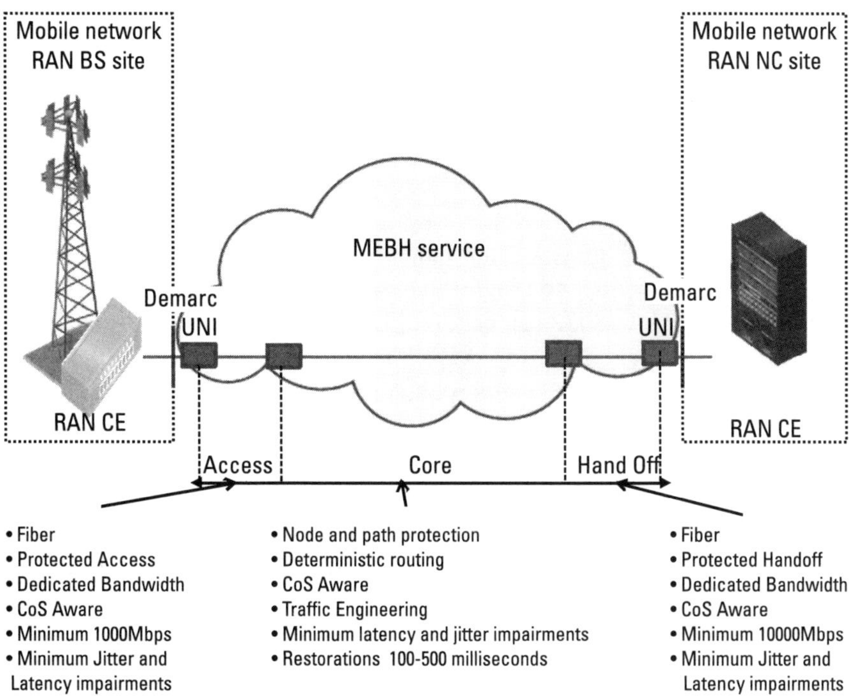

Figure 4.6 MEBH transport design.

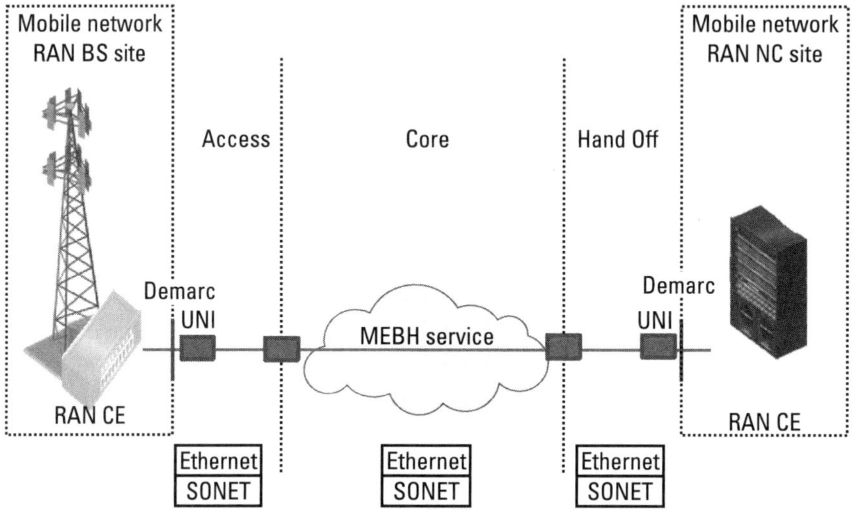

Figure 4.7 MEBH use case 1—SONET/SDH transport.

profiles due to mentioned overhead and inherent limited support for bursts may also be seen as the limitations for certain types of traffic.

Figure 4.8 presets the mixed transport solution. Access employs several technologies—fiber or SONET/SDH. Transport in core is based on the switches Ethernet (PBB), MPLS overlay, or SONET/SDH, and the handoff is using Ethernet over fiber technology. The design is fairly robust with good layer 1 protection in access and core if SONET/SDH is used (or OTU). The MPLS layer provides packet-level sophisticated protection and traffic engineering functions. An alternative use of PBB, in some services, may improve layer 2 features (domain scaling).

The use of fiber in access and handoff segments requires special protection design either by dual homing or the use of layer 2 protection technologies such as G.8031 [1], G.8032 [2], or LAG [3]; the fiber technology by itself does not provide protection. The use of SONET/SDH in the segment of the service will be the source of similar scaling problems and service logic limitations as in the SONET/SDH only design in use case 1.

Figure 4.9 presents the use case with different access technologies such as HFC, xDSL, and xPON. The functionality of core and handoff segments is similar to these in use case 2. The use of different access technologies may present a challenge due to the limitations of these technologies and the complex management of the service across different technologies. Yet, in some configurations, these technologies may be deployed extending the reach of service or lowering the access cost (using already existing infrastructure).

The best guide in the selection of transport is Table 3.9 with its comparison of service features supported by each of the transport and access technologies.

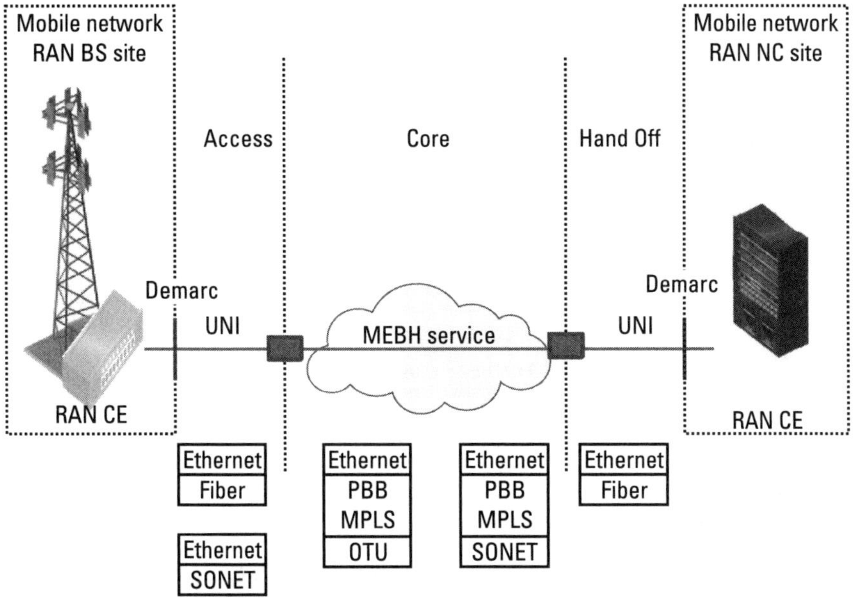

Figure 4.8 MEBH use case 2—mixed transport technologies.

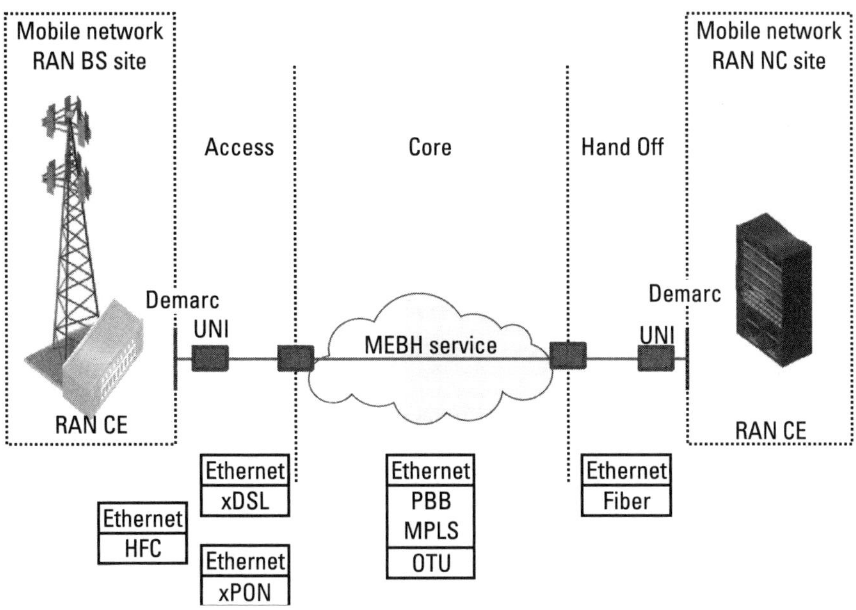

Figure 4.9 MEBH use case 3—mixed access technologies.

We could restate main principles guiding the design of the transport facilities for MEBH services as follows:

- The transport technology of choice should be mature but at the same time flexible offering the evolutionary path.
- The transport technology of choice should offer robust protection mechanism as the resiliency of the service will come in the most significant part from the transport technology.
- Mixing different transport technologies may result in operational complexities that may affect the service performance.
- The mantra of the MEBH services is growth, growth, and growth; the transport technology of choice should be able to support it.

4.3 MEBH Service Protection

If something can go wrong in the network, it will, given the reasonable amount of time.

—*The Tao of Network Design.*

The elusive objective of the MEBH service is 100 percent service availability. While nothing like 100 percent availability exists (even in science-fiction movies), wireless providers must come as close as possible to this imaginary [4] number if they want to stay in business. There are several ways to go about it and some rules to follow:

- The quality of the underlying network and transport is critical. The protection properties of the underling transport network must be understood and quantified.
- The logical design of the service has to provide redundant paths from the cell site to the MTSO. As explained before, the focus of the protection design should be on protecting facilities that carry aggregated traffic.
- The overall protection of the MEBH is a sum of protection offered by the logical design of the service and the resiliency of the MEBH network.

4.3.1 Protection Design

The reliability of service defined as availability is expressed in number of nines and segments these numbers apply to. There is an end-to-end availability, availability of transport and of handoff to the MTSO. The end-to-end availability of the MEBH will be a result of the service logic and the networking technology used by the service provider. One may, with the proper design, obtain higher service availability over a less reliable network. In principle, the reliable design requires backup facilities, whether these are physical resources or logical resources, such as dual homing toward the MTSO from separate SRNGs, or multiple EVCs in separate SRLGs. The examples of some possible designs with their reliability properties are provided in the previous sections.

Let's look now at three examples of the MEBH service design presented in Figure 4.10 from the perspective of service protection. Such an analysis may help to develop a similar "protection audit" for other designs.

Design A offers no facilities protection. It is a single EVC for each cell site with a single handoff to the MTSO. If this handoff fails, the service is lost for every EVC homing at this MTSO over this interface. The only protection in this design is provided by the design of the facilities transport in the MEN network. Of course, in this case it is of paramount importance to understand how MEN protection mechanisms are implemented.

Design B protects the handoff to MTSO and the MTSO itself. This is a reasonable choice for the design, as the MTSO handoff supports multiple EVCs and its failure would bring down large amount of services. Inside MEN,

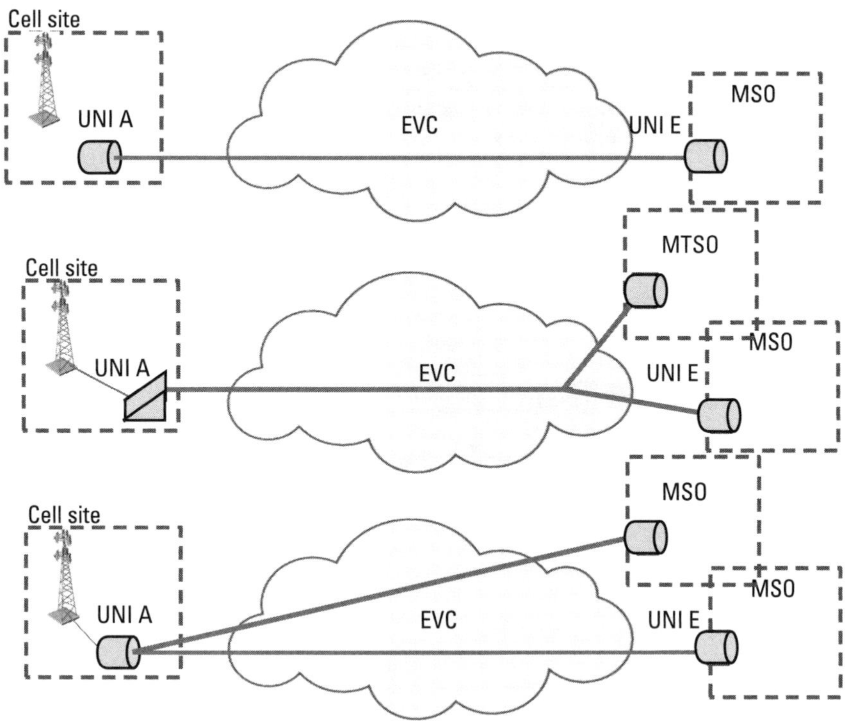

Figure 4.10 Protection designs for MEBH service.

however, the design protection relies on the MEN protection mechanisms only, as with case A.

Design C offers protection of the MTSO as well as of the transport from the ingress UNI. It is the most expensive of the three designs, but it offers the highest level of protection. Of course, the efficiency of it depends on the design of the MEN.

Designs A, B, and C do not have to be implemented with the EVPL service logic. One may experiment with different service logics to achieve a similar design as discussed in the service logic section.

4.3.2 SRG Analysis

For protection to work, the protected resources must be in the diverse shared risk groups (SRG). There is no point in building a protected logical design in which the EVCs that should be working as an active stand-by pair are placed in the same SRG. The analysis of resources subject to the same failures is a good exercise to go through; it may indicate the weakness in the design.

A word of caution: The analysis of the SRG is not clear-cut or easy, as the application to the actual design of the concept of SRG is open to different interpretations. Yet, the design decisions about which resources are shared and what are disjointed can be made relatively easy (not easy, but relatively easy!). The impact of other design elements, like the multilayer dependencies, are less obvious and require either abstraction of resources or detailed engineering analysis of the networking architecture. The example of the SRG analysis of the MEBH service is provided in Figure 4.11.

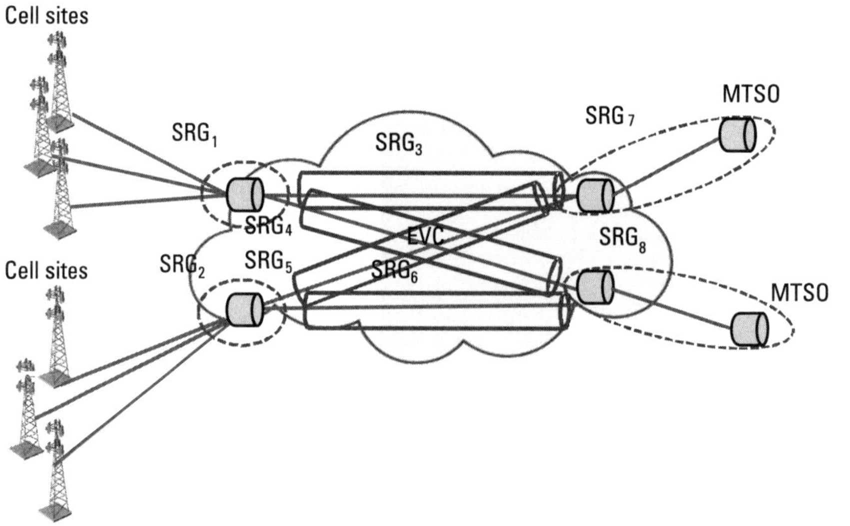

Figure 4.11 SRG analysis of MEBH service.

In the example we have eight SRGs: SRG1, SRG2, SRG3, SRG4, SRG5, SRG6, SRG7, and SRG8. SRGs 1 and 2 are access aggregation points. The failure of this facility would bring down all cell sites connected to it. One may in the design limit the number of UNIs per access SRNG. SRGs 3, 4, 5, and 6 are transport facilities in MEN. If the logical design of the service supports active and stand-by EVCs for the UNI, these should be placed in the disjointed SRG. One may, for example, limit the number of active EVCs per specific SRG. The handoff is in two SRGs, 7 and 8. The failure of one of them should not affect the other providing the protection for UNIs converging on both MTSOs. The limits on the amount of facilities in each SRG may be implemented by using multiple facilities (more than two MTSOs) and grouping the cell sites into regions not subject to the same risk factors.

A detailed audit of the MEBH service may discover how the actual implementation differs from the ideal design and suggest possible, or necessary, solutions. Alternatively, the SRG analysis of the service, if it will not force the changes in the design, at least may alert to the possible risks and risk areas inherent in the design.

As a final check on the reliability of the MEBH design, a model of the service can be developed and its reliability can be calculated using the method explained in Section 3.3.3. While the calculation does not give the exact availability score that one will see in the deployed service, it may help in eliminated some gross errors in the design, selecting alternative solutions or eliminating risky ones.

We could restate main principles guiding the design of the protected MEBH services as follows:

- The design should limit the number of resources per SRLGs, SRNGs, and SRDGs.
- The design should implement the service logic that would augment the protection design of the network.
- The design should distribute the critical service resources.

4.4 MEBH Quality of Service

The perception of the quality of the network services is inversely proportional to the amount of monitoring

—*The Tao of Network Design.*

The selection of the specific service logic should not have an impact on the QoS specifications; one needs to provide the required service quality regardless of the logic of connectivity. If with the selected service logic and transport, QoS requirements for MEBH service are somewhat compromised, the design needs to be revised. In defining the QoS aspects of the MEBH service, we need to define and establish requirements for the following:

- Bandwidth profile parameters;
- Policing and shaping;
- CoS class for data traffic and OAM traffic;
- A method for signaling of CoS for specific traffic flows.

4.4.1 Bandwidth

The MEBH service has very strict requirements regarding bandwidth available to the service. After all, the bandwidth is what the MEBH revolution is all about! The bandwidth is specified for EVC or UNI as a profile, as explained before, defining committed bandwidth (CIR) and burst capability (CBS) and EIR and EBS if required. CIR should be defined in terms of the layer 2 service frames, not layer 1, or layer 2 Ethernet frame with control fields.

CBS is another critical property of the QoS definition. The traffic from the cell site to the MTSO and back will never be uniformly distributed over time. It will come in bursts. It is due to the nature of the services as well as the type of equipment (some equipment cannot shape the traffic to the required bandwidth profile). If the service is not supporting necessary CBS parameters, it will never achieve CIR.

It is quite often difficult to conceptualize EIR/EBS bandwidth parameters in terms of the service. On one hand, they offer a promise of some traffic being transported. Yet, on the other hand, they carry the risk. So, what does the EIR/EBS mean from the practical point of view? We need to state clearly that these parameters offer no guarantees of service. The EIR/EBS bandwidth in all probability will be available at short time spans for bursty traffic patterns. One cannot count on them for the sustained traffic flows. What is more important, one cannot count on these resources during the high traffic demand (i.e., when resources are required the most). In these periods, only the committed bandwidth (CIR) will be available to the service. Thus, in sizing the required bandwidth for the MEBH service, one need to determine how much bandwidth one needs to support his service 100 percent of time. This is the required CIR parameter. Then, one may look at occasional bursts in traffic demand that may be coming from customers. This estimate would be the EIR and EBS parameter (the EIR/EBS is the bandwidth that is available when it isn't needed). The design may range from the design with CIR = 100 percent and EIR = 0 percent guaranteeing always the bandwidth to the design with some compromises like CIR = X percent and EIR = Y percent. Mixed designs will be less expensive, as the EIR/

EBS bandwidth is usually cheaper than CIR/CBS, and they will work 90 percent of the time or so. The designs with CIR = 100 percent of the demand while expensive would work (almost) always, barring failure conditions.

Another point: Not all CIRs are created equal. The CIR traffic in the high class (H) has better performance parameters than the CIR in the M traffic class. It is because the H traffic goes into the queue that is always serviced first (strict priority). The traffic in the M class is put into the queue that is serviced after the strict priority queue or queues are done (switches may have more than one strict priority queue). The way to conceptualize these two CIRs is to think about emergency vehicles. One with a red light and horn on gets priority on the road—this is H class traffic. The other is the same emergency vehicle but with the light and horn off. It will get through eventually, but the customer may be done by the time it arrives. Thus, one has to carefully plan which CIR traffic one wants to use in MEBH service.

4.4.2 Policing and Shaping

The preservation of the CIR and CBS (and other) parameters requires on the side of the service provider the proper policing. On the side of the customer, it may require shaping. The service provider may offer policing per EVC, per group of EVCs, or per specific CoS class. Policing per group of EVCs may offer an advantage of flexible bandwidth utilization missing with the per EVC (one level) policers. The parameters of the bandwidth—CIR or CBS—will impose requirements on the customer's equipment in terms of the required granularity of shaping and policing.

4.4.3 CoS Class Selection and Signaling

The service provider offers the service defined in specific classes of service. Each CoS class has its own properties expressed in measurable traffic characteristics and its specific signaling. Thus, the customer must select the class of traffic she wants to use and align her marking of CoS classes with that of the service provider, if the service provider cannot offer transparent CoS signaling. The previous sections on QoS define generic service CoS classes and their structure (CIR, CBS, EIR, EBS, metrics, and marking). The next section provides definitions of the LTE CoS classes and discusses the problem of mapping the LTE classes into the typical CoS classes offered by the MEBH providers. The customer may select one class that fits all. But this is usually an expensive service; as the CoS class chosen for this function has to satisfy the most stringent requirements. One may also select the multi-CoS service. The problem here, however, is that the customer does not know her traffic profile and in selecting multi-CoS service she may underestimate the need for the specific CoS profiles.

One may seek guidelines for the selection of the appropriate CoS parameters for the MEBH service in the LTE specifications. LTE service with nine defined classes of service (see Table 4.3) when interfacing with the MEBH service faces the CoS mapping problem; the Ethernet service defines at most eight CoS classes using the PCP bits in the VLAN tag header (CoS signaling methods based on EVC, EVC +PCP, and EVC +DSCP combinations have been defined but are not universally available as service options). The MEBH providers implement usually three or at most four CoS classes. Thus, potentially nine classes of traffic must be mapped into three or four classes in the MEBH service (see Table 3.25). There is no easy or obvious way to do it. Thus, it is no surprise that some wireless providers opt for a single, highest CoS class in MEBH service for all the traffic rather than multi-COS flow mapping that would involve complex CoS and bandwidth management (this is a situation in 2012). A single class of service is a simple option, but an expensive one. It may be expected that as the MEBH service matures it will support multiple CoS flows (two, three, or four), but certainly not nine. Thus, the CoS mapping problem between the LTE flows and MEBH service will stay around.

Note: The requirements from Table 4.3 do not translate into exactly the same requirements for the MEBH service. They are presented here to illustrate the type of, and the relative differences between, the classes of traffic in the LTE service.

In selecting the specific BW parameters one must consider the tradeoffs between the performance metrics achievable in each for the CoS classes and the traffic patterns. The CoS H class guarantees the lowest FDV and FTD targets. However, this class of traffic is very intolerant of bursts (large CBS). The CoS

Table 4.3
Standardized QoS Class Identifier (QCI) Characteristics for LTE Service [5]

QCI	Priority	Packet Delay Budget	Packet Error Loss Rate	Example Services
1	2	100 ms	10^{-2}	Conversational voice
2	4	150 ms	10^{-3}	Conversational video (live streaming)
3	3	50 ms	10^{-3}	Real-time gaming
4	5	300 ms	10^{-6}	Nonconversational video (buffered streaming)
5	1	100 ms	10^{-6}	IMS signaling
6	6	300 ms	10^{-6}	Video (buffered streaming) TCP-based (e.g., www, e-mail, chat, ftp, p2p file sharing, progressive video, etc.)
7	7	100 ms	10^{-3}	Voice, video (live streaming), interactive gaming
8	8	300 ms	10^{-6}	Video (buffered streaming), TCP-based (e.g., www, e-mail, chat, ftp, p2p file sharing, progressive video, etc.)
9	9	—	—	—

M class has less stringent FDV and FTD targets, but it is more tolerant of busty traffic [6]. One cannot increase the burst capacity of the H class and retain the same performance metric targets. The proper balancing of burst handling capacity and performance in selecting for the service traffic classes is one of the critical tasks facing the MEBH design.

We could restate main principles guiding the design of the QoS for MEBH services as follows:

- Define the bandwidth per EVC and the class of the traffic.
- Review policing and shaping requirements and capability of the CPE equipment.
- Define the signaling methods for the QoS class.
- Establish performance targets for the data and management traffic.

4.5 MEBH Service Performance

No network system will always behave as expected.

—The Tao of Network Design.

MEBH service performance objectives are defined by the performance budget of the wireless service. Using it as a guide, one must apportion to the MEBH segment appropriate targets for frame transfer delay, frame loss, frame delay variation, and availability and decide on the appropriate metrics for each of these measures.

As discussed previously, there are several methods of calculating each metric. Thus, to get reliable and comparable results one must provide the precise definition of the metric, including the way it is calculated.

One must be aware that the monitoring of metrics is usually done by a synthetic stream of test packets. The measurements collected with this method are only approximating the actual performance of the data flow [7]. There is no obvious and straightforward way to translate these measurements into the actual state of the service. One way of dealing with this indeterminacy is to build enough error margins that the measurements do not underestimate the performance of the service significantly.

A related aspect of the metric definition is the specification of the statistics by which the metric is reported. Any metric can be reported as specific statistics of the population of collected measurements over some interval. Typical ways of reporting metrics (e.g., frame transfer delay and frame delay variation) is to report a percentile from the measurements over one hour, one day, one week, or one month. Frame loss ratio is reported as the percentage of actual losses over the specific period. Measurements usually exclude periodic maintenance. One has to keep in mind that measuring the traffic performance is akin to statistical sampling rather than to the measuring a length of the piece of paper. Sampling results are always subject to inaccuracies that result from the sampling procedure and the unpredictable nature of the measured phenomena.

To characterize the MEBH service, one should select well-defined, well-understood metrics. It is a good practice to follow MEF or ITU guidelines on this. As well, the metrics should be appropriate to the task. For example, if part of the MEBH service includes packet-based synchronization, to verify such a service one should use metrics specific for this task and not approximate the service performance with generic metrics.

Performance objectives for the MEBH service are difficult to find. One reason for this is that the target FD, FDV, and FDR depend on wireless technology and disclosing them would somewhat expose proprietary information about the service. Instead of providing the actual target metric values we may, using the QoS classes defined in MEF, assign the type of traffic that MEBH is supporting to the specific class. Such a proposal was provided in Table 3.25.

The information in Table 3.25 should be used in the following way. In a case where the wireless operator uses a single CoS MEBH service, the performance metrics of the highest class must apply. If she is using two classes, the performance metrics of the highest and medium class would apply, respectively, and so on.

The performance objectives for the specific class (H, M, L) are usually defined by the type of applications supported in these classes. The bound on the metrics is established by the properties of the architecture (a physical system) on which the service is implemented. Table 4.4 provides the exemplar ranges of values of the four metrics in metro areas for the H CoS class. One can certainly design a service with metrics outside of these bounds, but in most cases one may expect to get the service within the metro area within these values.

We could restate main objectives guiding the design of the performance for MEBH services as follows:

Table 4.4
Exemplar Ranges of Performance Metrics in Metro Area in the H CoS Class

Metric	Symbol	Range	Unit
Frame delay one way	FTD	8–15	msec
Frame delivery ratio	FDR	99.9–99.999	%
Frame delay variance	FDV	1–5	msec
Availability	Av	99.95–99.999	%

- Define clearly performance metrics used to verify the performance of the service. The definition of the metric should go beyond the term only.
- Define target performance for each metric.

4.6 MEBH Service Verification

When a network fails, it tends to fail in places where it is not monitored.
—The Tao of Network Design.

The service that is not monitored usually performs to expectations. This seemingly paradoxical statement comes from the fact that if the service is not monitored, it is tacitly assumed that it runs as expected; there is no compelling reason to think otherwise. Only by monitoring the service may one uncover problems and issues. Thus, the necessary part of the service MEBH is a monitoring overlay. This overlay may be build using the SOAM functionality or the dedicated set of networking testing equipment.

4.6.1 Monitoring Overlay

One place to start the design of the service verification overlay is to clearly demark the segments of the service—define which segment of the network belongs to the provider and what belongs to the customer domain.

Figure 4.12 illustrates the element of the MEBH service and its ownership. The wireless provider looks at the service from UNI-C to UNI-C. The MEBH service provider owns the service from UNI-N to UNI-N. UNI-N is a part of UNI realized in the provider's network. UNI-C is a part of UNI realized in the customer's network. The MEBH service provider is responsible and will

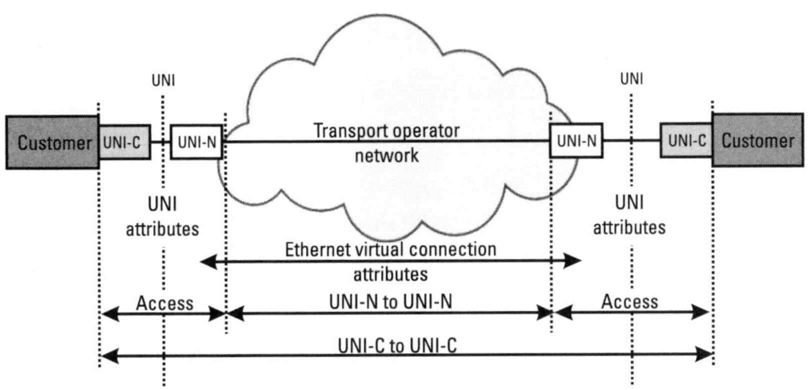

Figure 4.12 Network segments and service scope [8].

be monitoring the service between UNI-Ns. The customer of the service will be monitoring the service between UNI-Cs. There is of course no-man's land between them that the customer needs to be aware of. Figure 4.12 also illustrates the important point that the customer and the provider do not look at precisely the same segment of the network.

The example of the monitoring overlay with the dedicated testing devices is presented in Figure 4.13. The MEBH service provider monitors the service between his end testing equipment with the scope of the service. The wireless customer monitors the service from the location of his CPE devices. These two measurement approaches should in principle measure the same services, and in most cases they do. Yet, because of the different scope covered by these measurements, there may be cases when the measurement will differ. This is an important point to retain, as it is often the case that service problems reported by such monitoring systems may originate outside of the service provider's scope.

How the testing devices are connected to the service is the design question and is specific to the network and devices. Some network elements may have built-in the testing functions, thus eliminating the need for a separate testing device.

4.6.2 Monitoring Method

In defining the service verification requirements for MEBH service, one should define the method of collecting statistics including the method of measurements, the frequency of measurements, and location of monitoring devices if such are used. The service verification portion of the MEBH design should also include a list of all measurable service properties and their verification, not limited to performance metrics and including bandwidth parameters such as CIR,

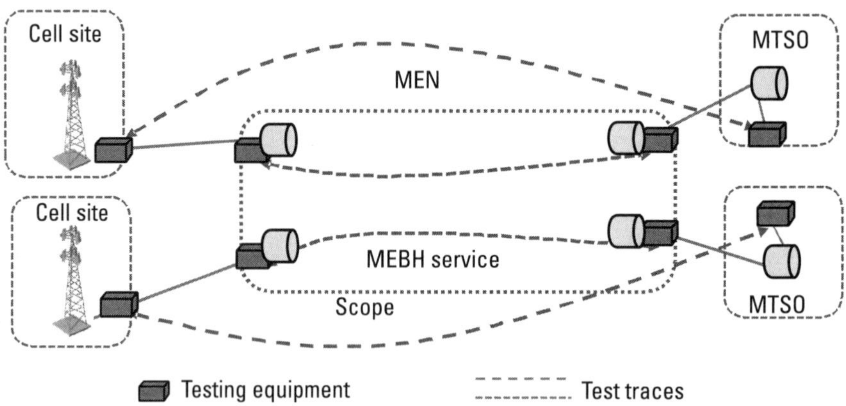

Figure 4.13 Performance monitoring overlay over MEBH service.

CBS, and if applicable EIR and EBS, as well as availability of the service and of specific elements of the service such as UNIs and EVCs.

Bandwidth parameters are usually verified only during the service activation process or during problems with the throughput. Thus, the service activation process should be also part of the service verification requirements. One may use here a service activation [10] process as defined in the ITU Y.1564 [11]. Any measurement specifications should define the scope of metrics including the demark points in the network determining the service segment over which metrics are defined. Last but not least one should define mechanisms for fault/status propagation/verification such as SOAM. Either IEEE (IEEE.802.1ag) or ITU (Y.1731) defined (as described in Section 3.6) algorithms could be used here.

4.6.3 SOAM Support

MEBH service providers should at the minimum support the transparent transport of the customer Y.1731 traffic. In some cases, the MEBH service provider may establish MIP points at the edges of its network that can help the customer to verify its connectivity. The internal SOAM architecture from the perspective of the MEBH customer should be transparent to his service.

Figure 3.53 may be difficult to translate into the network architecture. The concept of the SOAM services may be simplified as depicted in Figure 4.14. The design on this figure represents what most often the customer will need to understand from the SOAM architecture.

The design presents the basic components of the SOAM architecture deployed for the MEBH service. The design has three segments, or MEs: the ME for the mobile operator, the ME for the service provider, and the ME for the interface segment between the service provider and the mobile operator. Each ME is terminated with MEP, and the service provider on its UNI implements

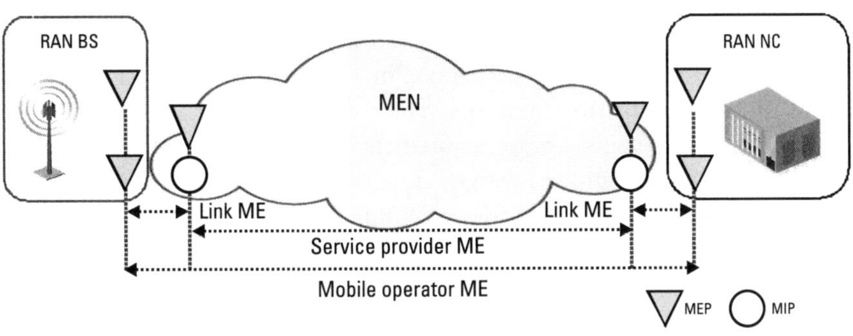

Figure 4.14 SOAM design in MEN [12].

MIP as well. This MIP can be used by the mobile operator to verify its connectivity to the service provider interface. The mobile operator SOAM traffic is transparent to the service provider. The status of the link ME may be signaled to the service provider ME and to the mobile operator. The mobile operator ME may support continuity checks as well as performance verification. The service operator SOAM design may include several lower MEs invisible to the mobile operator. Additional functionality may be included in the design such as loopbacks or fault management.

We could restate main principles guiding the design of the service verification for MEBH services as follows:

- Establish service verification architecture with a clearly defined scope.
- Establish SOAM architecture with reserved SOAM levels for your traffic; establish SOAM transparency rules and other SOAM services (LB).
- Establish a procedure for resolving discrepancies between performance measurements by the MEBH service provider and wireless customer.

4.7 MEBH Service—Interconnectivity

A network design should be feasible as well as reasonable.

—The Tao of Network Design.

The specification of interconnectivity requirements is relatively, at the stage of high-level architecture design, an item of low impact on the shape of the overall design. One still needs to define in some detail the transport media, the physical interfaces, and the physical specifications. In particular, if the service would restrict possible choices of technology (type of fiber, type of media, and so on). The interface requirements are well standardized, choices are well known, and they affect not the overall service design but the local implementation. The best approach is to use standards as explained in Section 3.7. The specific choices for the physical interconnectivity are defined in Table 3.30, Table 3.31, and Table 3.32. The providers usually support several configurations.

The physical (engineering) requirements are technology and site specific, so at the stage of the high-level design one may only indicate the desired choices and possible restrictions or limiting conditions.

We could restate main principles guiding the design of the engineering requirements for MEBH services as follows:

- Define any exclusions or restrictions on the design of service interconnectivity.

- Define preferred standards the interface should use or support.
- Define any requirements or restrictions on power feed, environment, location, dimensions, and site-specific conditions that may influence the design.

Endnotes

[1] The ITU-T G.8031 standard defines the APS protocol and linear protection switching mechanisms for point-to-point VLAN-based ETH subnetwork connection (SNC) in Ethernet transport networks defined in G.8031/Y.1342 Ethernet Linear Protection Switching standard (2010).

[2] The ITU-T G.8032 standard defines the Ethernet ring protection switching technology providing for sub-50-ms protection and recovery switching for Ethernet traffic in ring topology and loop prevention at layer 2 defined in ITU-T G.8032 Ethernet ring protection switching (2008).

[3] LAG, or link aggregation, is a technology used for link capacity augmentation or protection defined in IEEE 802.1ax standard.

[4] This is really an imaginary number, nothing to do with $\sqrt{-1}$.

[5] Based on Table 6.1.7 of 3rd Generation Partnership Project, Technical Specification Group Services and System Aspects, Policy and Charging Control Architecture (Release 8). 3GPP TS 23.203 V8.6.0 (2009-06).

[6] The business of the traffic in the MEBH service comes not only from the traffic patterns, but also from the ability (or inability) of some of the cell site equipment to shape the traffic to the desired level without introducing too much of delay or jitter.

[7] On the problems of the synthetic network performance measurements, see, for example: Sommers, J., et al., "Accurate and Efficient SLA Compliance Monitoring," SIGCOMM'07, August 27–31, 2007, Kyoto, Japan; Morton, A., et al., "New Network Layer Metrics for Packet Loss, Delay, Delay Variation," IETF 72 IPPM Future Work BOF, Dublin, Ireland, July 30, 2008; Xiaoyuan, T., *A Quality of Service Monitoring System for Service Level Agreement Verification*, M.E Thesis, The University of Sydney, 2006.

[8] Adapted from ITU T Y-1564, Ethernet Service Activation Test Methodology. ITU, 2011.

[9] These functions may support some performance monitoring protocols such as the TWAMP protocol defined in RFC 5355: A Two-Way Active Measurement Protocol (TWAMP), IETF October 2008.

[10] As defined by Y.1564, the service activation process covers "the service configuration test that tests each defined Ethernet service to make sure that the configuration is correct. Each of the important parameters (IR, FTD, FDV and FLR) *must* be tested. This test is very short in length at the judgment of the person performing the test and is designed to prevent wasted time caused by failed service performance tests. And the service performance test is conducted to validate the quality of the Ethernet services over a medium to long time duration." ITU- T Y.1564.

[11] ITU-T Y.1564, Ethernet Service Activation Test Methodology, 2011.

[12] Olsson, J., "Mobile Backhaul," MEF presentation, May 2009.

5

MEBH Service: A Use Case

While designing a network for the future is difficult, designing for today is disastrous.

—*The Tao of Network Design.*

5.1 Use Case Assumptions

We are now ready to apply the discussed service concepts to the design case. We should preamble this design exercise with a note seen on several movies that try to depict some real-life situations that all similarities of the proposed design to real-life implementations or services are purely accidental only, residing in the eyes of the beholder. To be more explicit, we need to indicate the values/parameters in the use case are for illustrative purposes only and should not be interpreted as requirements for MEBH.

The exemplar MEBH service that we will be studying requires 50 Mbps of the committed (CIR) bandwidth (measured at layer 2) to each cell site with the possibility of a migration to bandwidth above 150 Mbps; the bandwidth is symmetrical. In addition to the CIR guarantees, the services need to support up to 128 KB committed burst size (CBS). In most cases, the service will use one QoS class, the class H, in which the total contracted bandwidth is guaranteed with the highest SLA performance targets. The mutli-QoS design should, however, also be feasible at the later stage of the service deployment. This condition means that the design implemented initially should not preclude the deployment of the multi-CoS service.

Few service features like the intercell-site channel to support the X2 interface (about 10 percent of the data channel capacity in class M) and the separate dedicated management channel between the cell sites and the MTSO (5–10

Mbps in class M) are optional. The handoff to the MTSO should be protected. The traffic from cell sites to the MTSO should be grouped into the SRGs by n cell sites in the group. The end-to-end service availability should be close to four nines, and the handoff to MTSO should support the availability above four nines. The recovery from node or link failures on protected segments should be in the range of 300–600 milliseconds. Three metrics—frame transfer delay, frame loss ratio, frame delay variation—should be monitored and reported for the violations of the specific performance targets. The use of the network availability metric is optional but desired. The service should use SOAM overlay with CCM at levels 5–6 and be transparent to the MEBH service. The service may offer LB points at the edges of the MEBH service. The handoff to the cell site CPE and to the MTSO CPE has to be over fiber. No copper conduit is to be included in the path. The UNI handoff at the MTSO is 10 Gbps; the cell site should be 1 Gbps. All other aspects of the service design depend on the specific service construct. Table 5.1 summarizes these service assumptions. Figure 5.1 presents the conceptual design for the discussed use case.

We select the service models based on the EVPL service logic. For protection of the service each cell site will be homing to two MTSOs. The service is augmented with the EVP-LAN EVC that would provide an additional channel for management traffic. This EVC will also serve for the intercell site communication if required. L2CP rules will follow the MEF 6.1.1 requirements. Figure 5.2 presents the service logic design for the use case and Table 5.2 presents the exemplar EVC UNI map.

Table 5.1

Common Assumptions Behind Exemplar MEBH Service Models

Service Feature	Status	Attributes
UNI at cell site	Required by each service	Supporting CIR of 50 Mbps, possible migration to 150 Mbps; fiber interface
UNI at MTSO	Required by each service	10 Gbps; fiber interface
QoS	Required by each service	H class of service; CIR= 50 Mbps, CBS = 128 KB; SLAs guarantees on contracted BW
mQoS	Optional	H and M QoS class
Intercell-site channel	Optional	10 Mbps; CoS M class
Dedicated management channel	Optional	10 Mbps; CoS M
Protection design	Required	Reliability four nines end to end and greater than four nines on handoff to MTSO; restoration times in the range of 300–600 milliseconds

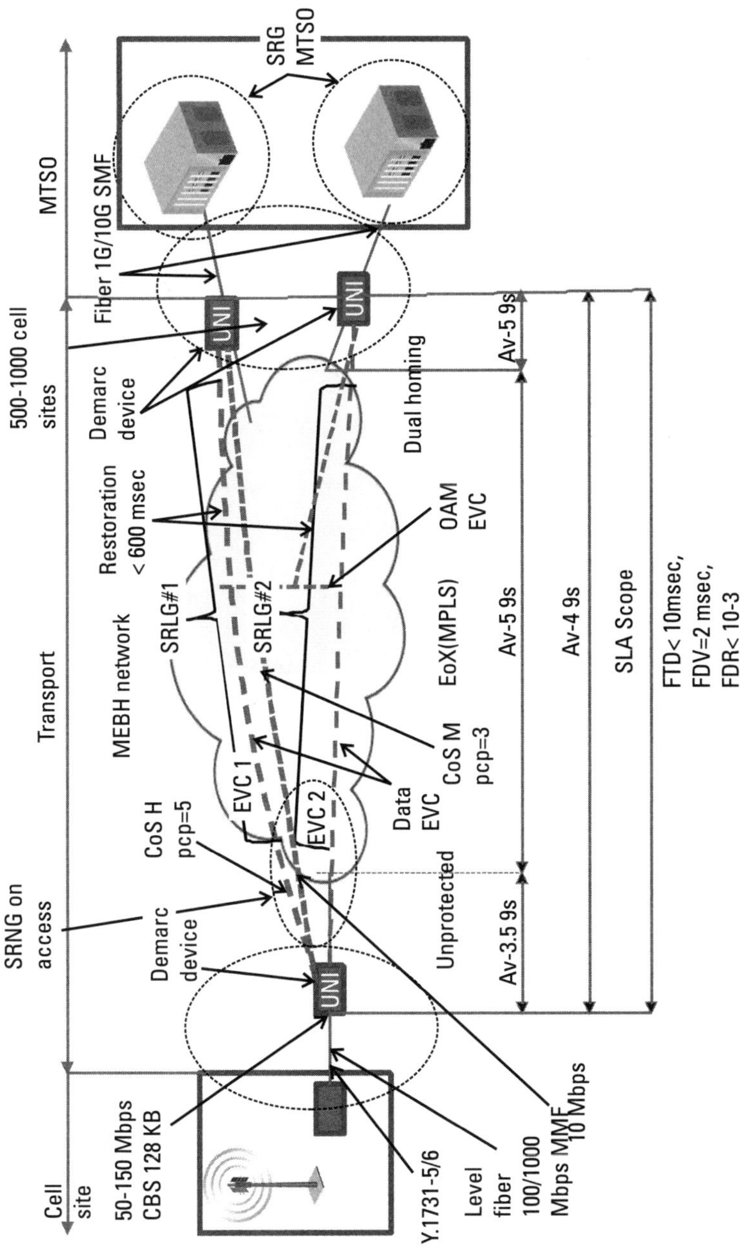

Figure 5.1 Conceptual design for MEBH use case.

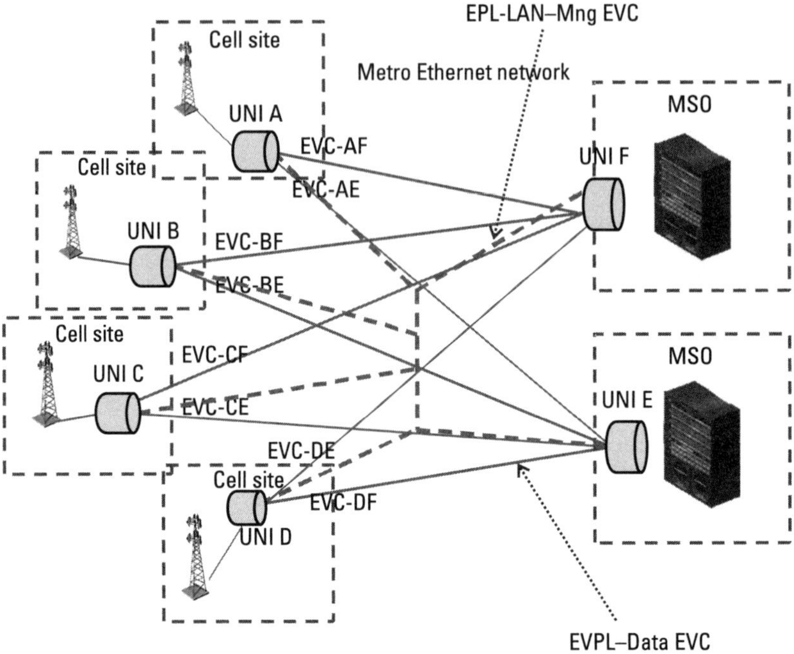

Figure 5.2 Service logic for MEBH use case.

Table 5.2
MEBH Use Case—EVC UNI Map

EVC-AF	UNI A; UNI F
EVC-AE	UNI A; UNI E
EVC–EVP-LAN	UNI A, UNI B; UNI C; UNI D; UNI E; UNI F

5.2 Use Case Requirements

From the relatively unstructured description of the MEBH service presented here, we construct the set of service functional requirements. These are presented in Table 5.3 [1]. These are example specifications, not signifying any specific equipment or service.

5.3 Use Case EVC UNI Attribute Table

Finally, we can come to the EVC and UNI attribute Table 5.4 and QoS profile in Table 5.5.

Table 5.3
Service Requirements for MEBH Use Case

Functional Group	Requirements (MEBH Service Use Case)	Design
Service logic	EVC type	EVPL and EVP-LAN
	Number of EVCs on UNI cell site	3
	Number of EVCs at MTSO	500–1000
	BUM processing	Must forward and filter
	Reserved VLAN IDs*	1 and > 4090
	L2CP at UNIs	MEF 6.1.1
	Processing of untagged frames	Drop
	VLAN IDs	Any in 2–4090; MTSO scope
Transport	Access technology requirements	Fiber, SONET, or DWDM, no wireless, no GPON, no DSL
	Transport technology requirements	MPLS or OTU
	MTU	≥ 2000B
Protection	Target availability for EVC, UNI, and MTSO UNI	≤ four nines end to end; ≥ four nines at MTSO; 3.5 nines at cell site
	Convergence times for different failure modes	< 300–600 milliseconds in transport segment
	Shared risk node or link groups†	Cell site access on SRNG < n cells per SRG
	Transport design protection aspects‡	EVCs from cell site in separate SRLGs; UNIS to MTSO in separate SRG link and node
Quality of service	Bandwidth required and bandwidth required and UNI	50–150 Mbps CIR; 128-KB CBS in class H; CIR 10 Mbps in class M; CIR, CBS, EIR, EBS
	Signaling of class of service	Pcp class H; pcp = 5; EVPL class M; pcp = 3; EVP-LAN pcp bit transparency
	Policing or shaping requirements	Per EVC ingress policing; color-blind; UNI CAC§ 80–90 percent of UNI speed
Service performance	Metrics for each class of service	For H class: FTD < 10 msec; FDV < 2 msec; FLR < 99.99 percent For M class: FTD < 20 msec; FDV < 8 msec; FLR < 99.9 percent
Service verification	SOAM architecture	Y.1731 transparency for level 5 and 6; LB availability
	SLA measurements methodology	Per EVC, per CoS, up to 1 to 10 pps synthetic test frame traffic
	Metrics reporting methodology	Metrics reported per month; reported statistics 95–99 percent percentile for FDT, FDV; FDR reported per month per EVC
	Fault notification	Using Y.1731

Table 5.3 (continued)

Functional Group	Requirements (MEBH Service Use Case)	Design
Interconnection—engineering specifications	Types of permissible interfaces at UNIs	1G and 10G IEEE 802.3-2008 Section 3 fiber standards, no copper
	Required fiber types at UNIs	MMF and SMF
	Space, environment requirements	Environmental hardened; CO certified; temperature range: –40°C (–40°F) to 60°C (140°F)
	Permissible power feed types	+24 VDC and –48 VDC

Notes:

* The reason for reserved VLANS VIDs is that quite often the provider, if he is not isolating his traffic from the customer traffic, uses some VIDs for his internal network management. The mark of good service is a complete isolation of the customer's VLAN ID schema from that of the provider of the service. Thus, the customer should not worry about any internal provider VLAN restrictions.

† The number *n* of facilities per SRG is not a set standard but depends on the desired architecture and protection. Quite often, it is difficult to implement such a requirement. But it is good to know in any case what the actual n deployed is.

‡ Implementing the separate SRGs per EVCs and handoffs is critical to the resiliency of the design and service.

§ The BW is defined in terms of layer 2 service frames without overhead and control traffic (like PM). That is, it's a good practice to budget the UNI CIR with some room left.

Do these specifications exhaust the entire design requirement? Of course, they do not. However, they represent the core of specifications around which the more complete design of service can be built.

5.4 Conclusions and Final Thoughts

This book was arranged in a way that would guide someone involved in the design of the MEBH architecture through the necessary steps to design a workable framework for the service. Yet, some ambiguities may still be present and some connections between the components of the design may have escaped the attention of the reader. Of course, the design process is not and cannot be mechanized, even when given a clear guideline; one still may make incorrect choices. Even if the design process in the book was organized to provide the logical progression from basic concepts to the final design, many aspects of the process and design details have been omitted for lack of space. Thus, a lot is left unsaid potentially inviting poor design decisions. However, no book may prevent this from happening.

Table 5.4
MEBH UC EVC and UNI Attributes

Object Attributes	Service Attribute	EVC	Cell Site UNI	MTSO UNI
UNI attributes	Physical medium	NA	Ethernet	Ethernet
	Speed		1 G	10G
	Mode		Duplex	Duplex
	MAC layer		802.3	802.3
	UNI MTU size		Greater than 2000	Greater than 2000
	Service multiplexing		Yes	Yes
	Bundling		No	No
	All-to-one bundling		No	No
	CE-VLAN ID for untagged and priority tagged service frames		None	None
	Maximum number of EVCs per UNI		>3	> 500
	Ingress bandwidth profile per UNI		EVPL CIR, CBS	EVPL CIR, CBS
	Egress bandwidth profile per UNI		None	None
	Layer 2 control protocols processing		MEF 6.1.1	MEF 6.1.1
EVC attributes	EVC type	Site specific	NA	
	UNI list	Area specific		
	Maximum number of UNIs	2 on EVPL; 500 on EPL-LAN		
	EVC MTU	≥ 2000		
	CE-VLAN ID preservation	Yes		
	BMU frame delivery	Unconditional		
	L2CP	MEF 6.1.1		
	EVC performance	As per SLAs		
	CE-VLAN COS preservation	Yes		
	CE-VLAN ID/EVC map	NA	Site specific	Site specific
	Ingress bandwidth profile per EVC		Yes	Yes
	Ingress bandwidth profile per CoS ID		No	No
	Egress bandwidth profile per EVC		None	None
	Egress bandwidth profile per CoS ID		None	None

Table 5.5
QoS Profile for the MEBH Use Case

CoS Label	Frame Transfer Delay (FTD)	Frame Delay Variation (FDV)	Frame Loss Ratio (FLR)	Availability (Av)	Bandwidth Profile Parameters	PCP CoS Coding
H	10 ms	2 ms	10–4	Four nines	CIR=50Mbps, CBS=128KB	5
M	20 ms	8 ms	10–3	Three nines	CIR=5KB EIR=10Mbps, EBS=20KB	2–3

Endnote

[1] All requirements in Table 5.3 represent exemplar requirements, not representing any particular service or service technology or vendor preferences.

Appendix A
Ethernet Standards

Metro Ethernet Forum (MEF) [1]

- MEF 2. Requirements and Framework for Ethernet Service Protection. MEF 2003. The document defines a broad framework for hop-by-hop and end-to-end service level protection.
- MEF 3. Circuit Emulation Service Definitions, Framework and Requirements in Metro Ethernet Networks. MEF 2004. The document defines the requirements for circuit emulation service "tunnels" TDM traffic through a metro Ethernet network allowing inclusion of legacy networks within a carrier Ethernet environment.
- MEF 4. Metro Ethernet Network Architecture Framework Part 1: Generic Framework. MEF 2004. The document introduces the framework and terminology for the services (Eth) layer and provides the fundamental understanding of the carrier Ethernet architecture.
- MEF 6.1 Metro Ethernet Services Definitions Phase 2. MEF 2008. The document defines service types (E-line, E-LAN, E-tree) and standardizes a few services based on the service types (EPL, EVPL, EP-LAN, EVP-LAN, EP-tree, EVP-tree).
- MEF 6.1.1 Layer 2 Control Protocol Handling Amendment to MEF 6.1. MEF 2012. The document aligns layer 2 control protocol treatment at MEF compliant UNI to be consistent with IEEE specifications.

- MEF 7.1 Phase 2 EMS-NMS Information Model. MEF 2011. The document provides a standard for carrier management systems to enable configuration and fault management of metro Ethernet services.
- MEF 7.1.1 Technical Corrections to MEF 7.1.
- MEF 8 Implementation Agreement for the Emulation of PDH Circuits over Metro Ethernet Networks. MEF 2004. The document gives precise instructions for implementing interoperable CES equipment that reliably transport TDM circuits across metro Ethernet networks while meeting the required performance of circuit-emulated TDM services as defined in ITU-T and ANSI TDM standards.
- MEF 9 Abstract Test Suite for Ethernet Services at the UNI. MEF 2009. The document defines the test suite for conformance of Ethernet services and equipment when deployed at the UNI. Specifically, the standard supports compliance testing of MEF 6, 10, 11, and specified required attributes.
- MEF 10.2 Ethernet Services Attributes Phase 2. MEF 2009. The document defines the service attributes and parameters required to offer the services defined in MEF 6.1. Updated from original MEF 10 and 10.1.
- MEF 10.2.1 Performance Attributes Amendment to MEF 10.2. MEF 2011. The document redefines the service performance parameters concerning availability, resiliency, and handle-related issues. Modifies specific sections in 10.2.
- MEF 11 User Network Interface (UNI) Requirements and Framework. MEF 2004. The document defines a split demarcation function between the customer (subscriber) and the service provider.
- MEF 12.1 Carrier Ethernet Network Architecture Framework Part 2: Ethernet Services Layer—Basic Elements. MEF 2010. The document defines the Ethernet services (ETH) layer as the specific layer network responsible for delivery of Ethernet protocol data units across internal and external interfaces as introduced in Table A.1.
- MEF 12.1.1 Carrier Ethernet Network Architecture Framework. Part 2: Ethernet Services Layer—External Interface Extensions. MEF 2011.
- MEF 13 User Network Interface (UNI) Type 1 Implementation Agreement. MEF 2005. This document defines the UNI attributes that allow existing Ethernet devices (switch, router, workstation, and so on) acting as customer edge devices to be compliant with this IA with no additional software or hardware upgrades.

- MEF 14 Abstract Test Suite for Traffic Management Phase 1. 2005. The document defines the requirements and corresponding test procedures for service performance and bandwidth profile service attributes that may be specified as part of a service level specification (SLS) for an Ethernet service. These tests are an essential part of the deployment of carrier Ethernet services since they provide a baseline for confident deployment of equipment that has already been seen to be compliant. This in turn greatly minimizes interoperability issues.
- MEF 15 Requirements for Management of Metro Ethernet Phase 1 Network Elements. MEF 2005. The document specifies the network management requirements to be met by network elements supporting Ethernet service phase 1.
- MEF 16 Ethernet Local Management Interface. MEF 2006. The E-LMI protocol specified in MEF 16 enables customer equipment to receive information regarding the status and attributes of Ethernet services, thus allowing automatic configuration and improved subscriber network performance.
- MEF 17 Service OAM Framework and Requirements. MEF 2007. The document provides requirements to be satisfied by the service OAM mechanisms in MENs and framework for discussing and implementing those mechanisms. It also provides context for several MEF specifications (UNI type 2 and E-NNI) and the work of other standards bodies.
- MEF 18 Abstract Test Suite for Circuit Emulation Services. MEF 2007. The document specifies testing procedures for pass/fail assessment of conformance with each of the operating modes in MEF 8.
- MEF 19 Abstract Test Suite for UNI Type 1. MEF 2007. The document supplements the MEF test specifications MEF 9 and MEF 14 with test procedures for UNI manual configuration mode defined in MEF 13.
- MEF 20 UNI Type 2 Implementation Agreement. MEF 2008. The document specifies MEF UNI characteristics and operation in which the customer side of the UNI is automatically configured by the network side of the UNI allowing verification of SLA and UNI connectivity. Additional objectives include support for Ethernet OAM (802.3ah, 802.1ag) over the UNI.
- MEF 21 Abstract Test Suite for UNI Type 2 Part 1 Link OAM. MEF 2008. The document provides the first of six possible test suites for UNI type 2 (MEF 20).
- MEF 22.1 Mobile Backhaul Phase 2 Implementation Agreement. MEF 2012. This document identifies the requirements for MEF Ethernet ser-

vices and MEF external interfaces (EIs such as UNIs) for use in mobile backhaul networks based on MEF specifications. In addition, new interface and service attributes have been specified where needed.

- MEF 23.1 Class of Service Phase 2 Implementation Agreement. MEF 2012. The document specifies a set of 3 Class of service names called CoS labels that can be used by operators, service providers and their subscribers to indicate the performance expectations to be associated with a given set of frames that comprise a CoS frame set. This CoS IA includes standards for CoS and color identification as well as performance objectives and supporting requirements. The CoS labels are envisioned as a subset of all of the class of service names an operator may provide
- MEF 24 Abstract Test Suite for UNI Type 2 Part 2 E-LMI. MEF 2009.
- MEF 25 Abstract Test Suite for UNI Type 2 Part 3 Service OAM. MEF 2009.
- MEF 26.1 External Network Network Interface (ENNI)—Phase 2. MEF 2012. This technical specification extends the ENNI (defined in MEF26) by defining the UNI tunnel access (UTA), which associates a virtual UNI (VUNI), a remote UNI, and at least one supporting OVC. MEF 26 specifies the reference point that is the interface between two metro Ethernet networks (MENs) where each operator MEN is under the control of a distinct administration authority. The ENNI is intended to support the extension of Ethernet services across multiple operator MENs.
- MEF 27 Abstract Test Suite For UNI Type 2 Part 5: Enhanced UNI Attributes & Part 6: L2CP Handling. MEF 2010.
- MEF 28 External Network Network Interface (ENNI) Support for UNI Tunnel Access and Virtual UNI. MEF 2010. This document extends the ENNI by defining the UNI tunnel access (UTA), which associates a virtual UNI (VUNI), a remote UNI, and at least one supporting OVC.
- MEF 29 Ethernet Services Constructs. MEF 2010. This document defines an abstract set of Ethernet service constructs that are widely applicable regardless of the UNI-to-UNI nsubnetworks or interfaces can do so by referring to and building upon this specification.
- MEF 30 Service OAM Fault Management Implementation Agreement.2011. This document specifies an implementation agreement (IA) for service operations, administration, and maintenance (OAM) that builds upon the fault management (FM) framework and requirements specified by MEF 17.

- MEF 31 Service OAM Fault Management Definition of Managed Objects. MEF 2011. This document specifies the fault management (FM) management information base (MIB) necessary to implement the service operations, administration, and maintenance (OAM) that satisfies the service OAM requirements and framework specified by MEF 17 [8], the service OAM fault management requirements as specified by SOAM-FM [10], and the service OAM management objects as specified by MEF 7.1, which are applicable to fault management functions.
- MEF 31.0.1 Amendment to Service OAM SNMP MIB for Fault Management. MEF 2012.
- MEF 32 Requirements for Service Protection Across External Interfaces. MEF 2011. This specification identifies resiliency requirements at external interfaces applicable to MEF service and associated service end points, which associate services with external interfaces. MEF would like these requirements to be considered when designing a mechanism for service protection across an external interface.
- MEF 33 Ethernet Access Services Definition. MEF 2012. This document defines Ethernet access services, which are OVC-based Ethernet services in contrast to the EVC-based services, which are defined in MEF 6.1 Technical Specification "Ethernet Services Definitions." Specifically, this document defines the UNI, OVC, OVC per UNI, OVC end point per ENNI, and ENNI requirements for point-to-point OVC-based Ethernet services. In addition, an informative appendix is provided showing use cases of some of the defined services.
- MEF 34 ATS for Ethernet Access Services. MEF 2012.
- MEF 35 Service OAM Performance Monitoring Implementation Agreement. MEF 2012. This document specifies an implementation agreement (IA) for service operations, administration, and maintenance (SOAM) that satisfies and extends the performance monitoring (PM), framework, and requirements described in MEF 17.
- MEF 36 Service OAM SNMP MIB for Performance Monitoring. MEF 2012. This document specifies the performance monitoring (PM) management information base (MIB) necessary to manage service operations, administration, and maintenance (OAM) implementations that satisfy the service OAM requirements and framework specified by MEF 17, the service OAM performance monitoring requirements as specified by SOAM-PM, and the service OAM management objects as specified by MEF 7.1, which are applicable to performance monitoring functions.
- MEF 37 Abstract Test Suite for ENNI. MEF 2012.

- MEF 38 Service OAM Fault Management YANG. MEF 2012. This document specifies the performance monitoring (PM) YANG module necessary to implement the service operations, administration, and maintenance (OAM) that satisfies the service OAM requirements and framework specified by MEF 17 [9], the service OAM performance monitoring requirements as specified by SOAM-PM [13], and the service OAM management objects as specified by MEF 7.1 and ITU-T Q.840.1, which are applicable to performance-monitoring functions.

Institute of Electrical and Electronic Engineers (IEEE) [2]

- IEEE 802.1D-2004. MAC Bridges (rollup of 802.1D-1998, 802.1t, 802.1w, P802.1y, and 802.11c). IEEE, 2004.
- IEEE 802.1Q-2011. VLAN Bridges (Rollup of 802.1Q-2005+Cor-1 and 802.1ad/ag/ah/aj/ak/ap/Qat/Qav/Qaw/Qay/Qau). IEEE 2011.
- IEEE 802.1Qaz-2011. Enhanced Transmission Selection for Bandwidth Sharing Between Traffic Classes. IEEE 2011.
- IEEE 802.1Qbb-2011. Priority-Based Flow Control. IEEE 2011.
- IEEE 802.1Qbc-2011. Provider Bridging—Remote Customer Service Interface. IEEE 2011.
- IEEE 802.1Qbe-2011. Multiple Backbone Service Instance Identifier (I-SID) Registration Protocol (MIRP). IEEE 2011.
- IEEE 802.1Qbf-2011. PBB-TE Infrastructure Segment Protection. IEEE 2011.
- IEEE 802.1Qbg-2012. Edge Virtual Bridging. IEEE 2012.
- IEEE 802.1X-2010. Port-Based Network Access Control (revision of 802.1X-2004, including P802.1af). IEEE 2010.
- IEEE 802.1AB-2009. Station and Media Access Control Connectivity Discovery (LLDP) (revision to 802.1AB-2005). IEEE 2009.
- IEEE 802.1AE-2006. MAC Security. IEEE 2006.
- IEEE 802.1AR-2009. Secure Device Identity (DevID). IEEE 2009.
- IEEE 802.1AS-2011. Timing and Synchronization for Time-Sensitive Applications in Bridged Local Area Networks. IEEE 2011.
- IEEE 802.1AX-2008. Link Aggregation (Initially created as 802.3ad-2000). IEEE 2011.

- IEEE 802.3-2008 [3], IEEE Standard for Information Technology—Telecommunications and Information Exchange Between Systems. IEEE standard for local and metropolitan area networks, five books (sections) published Dec. 26, 2008.

Section	Content
Section 1; Clause 1 to 20; Annex A to H, 4A	CSMA/CD overview; MAC; PLS/AUI; 10BASE5 MAU; 10BASE2 MAU; 10BROAD36 MAU; 10BASE-T MAU; 10BASE-F MAUs; 10 Mbps repeater; 10 Mbps topology; 1BASE5; DTE & MAU management; repeater management
Section 2; Clause 21 to 33; Annex 22A to 33E	100 Mbps Overview; 100BASE-T2; 100BASE-T4; 100BASE-TX; 100BASE-FX; 100 Mbps repeater; 100Mbps topology; MAC control; auto-negotiation; management; DTE power; maintenance 6–8; 802.3-2005/Cor 1
Section 3; Clause 34 to 43; Annex 36A to 43C	1000 Mbps overview; GMII ; 1000BASE-X AutoNeg; 1000BASE-SX; 1000BASE-LX; 1000BASE-CX; 1000BASE-T; 1000 Mbps repeater; 1000 Mbps topology; maintenance 6–8
Section 4; Clause 44 to 55; Annex 44A to 55B	10 Gbps overview; MDC/MDIO; XGMII ;XAUI; XSBI ;10GBASE-SR; 10GBASE-LR; 10GBASE-ER; 10GBASE-SW; 10GBASE-LW; 10GBASE-EW; 10GBASE-LX4; 10GBASE-CX4; 10GBASE-T; maintenance 7–8 802.3-2005/Cor 2
Section 5; Clause 56 to 74; Annex 58A to 74A SA = subscriber access networks; BPE = backplane Ethernet	SA overview; OAM; MPMC; 100BASE-LX10; 100BASE-BX10; 1000BASE-LX10; 1000BASE-BX10; 1000BASE-PX10; 1000BASE-PX20; 10PASS-TS; 2BASE-TL; SA topology; BPE overview; 10GBASE-LRM; 1000BASE-KX; 10GBASE-KX4; 10GBASE-KR; BPE AutpNeg; 10GBASE-R FEC; maintenance 8

International Telecommunication Union (ITU)

- ITU-T G.8010/Y.1306 (2004), Architecture of Ethernet layer networks, plus Amendment 1 (2006) and Amendment 2. ITU-T, 2010.
- ITU-T G.8011/Y.1307 (2009), Ethernet service framework. ITU-T, 2009.
- ITU-T G.8011.1/Y.1307.1 (2010), Ethernet private line service. ITU-T, 2010.
- ITU-T G.8011.2/Y.1307.2 (2009), Ethernet virtual private line service. ITU- T, 2009.
- ITU-T G.8012/Y.1308(2004), Ethernet UNI and Ethernet NNI, plus Amendement 1. ITU-T, 2006.
- ITU-T Y.1563 (2009), Ethernet frame transfer and availability performance, plus Amendment 1. ITU-T, 2009.
- ITU-T Y.1564 (2011) Ethernet service activation test methodology. ITU-T, 2011.

- ITU-T Y.1730 (2004), Requirements for OAM functions in Ethernet-based networks and Ethernet services. ITU-T, 2004.
- ITU-T Y.1731 (2008), OAM functions and mechanisms for Ethernet based networks, plus Amendment 1. ITU-T, 2010.

Endnotes

[1] MEF documents are available at http://metroethernetforum.org/page_loader.php?p_id=29. A short summary of each document's content is available on the specific MEF document.

[2] Based on http://www.ieee802.org/1/ and http://www.ieee802.org/3. Accessed on November 20, 2012.

[3] IEEE 802.3, State of the Standard, IEEE 802.3 Working Group, San Antonio, Texas, November 12, 2012.

Acronyms

3GPP	Third generation mobile system
3GPP2	Third Generation Partnership Project
ADSL	Asymmetric digital subscriber line
AIS	Alarm indicator signal
APS	Automatic protection switching
ATS	Abstract test suite
Av	Availability
BH	Backhauling
BMU	Broadcast, multicast, unknown unicast
BW	Bandwidth
CCM	Continuity check message
CDMA	Code division multiple access
CE	Customer edge, customer equipment
CF	Coupling flag
CFI	Canonical form indicator
CFM	Connectivity Fault management
CM	Color mode
CN	Core network
CPE	Customer premises equipment

CRC	Cyclic redundancy check
DMM	Delay measurement message
DMR	Delay measurement reply
DOCSIS	Data over cable service interface specification
DSL	Digital subscriber line
DST	Destination
eNB	Evolved node B
ENNI	External NNI
EoC	Ethernet over copper
EoDWDM	Ethernet over DWDM
EoF	Ethernet over fiber
EoHFC	Ethernet over HFC
EoMPLS	Ethernet over MPLS
EoOTU	Ethernet over OTU
EoPON	Ethernet over PON
EoS	Ethernet over SONET
EoTDM	Ethernet over TDM
EoWDM	Ethernet over WDM
EPC	Evolved packet core
EP-LAN	Ethernet private LAN
EPS	Evolved packet system
E-UTRAN	Evolved TMTS terrestrial radio access network
EVC	Ethernet virtual connection
EVDO	Evolution data only/evolution data optimized
EVPL	Ethernet virtual private line
EVP-LAN	Ethernet virtual private LAN
ETR	Effective throughput rate
FCS	Frame check sequence
FDV	Frame delay variation

FER	Frame error ratio
FLR	Frame loss ratio
FLR	Frame loss ratio
FSC	Frame check sequence
FTD	Frame transfer delay
FTDV	Frame transfer delay variation
GFP	Generic framing procedure
GPON	Gigabit PON
GSM	Global System for Mobile Communication
HFC	Hybrid fiber-coaxial
HSPA	High-speed packet access
IEEE	Institute of Electrical and Electronics Engineers
IETF	Internet Engineering Task Force
IFG	Interfere gap
IP	Internet protocol
ITU	International Telecommunications Union
L2CP	Layer 2 control protocol
LAN	Local area network
LB	Loopback
LBM	Loopback message
LBR	Loopback reply
LLC	Logical link control
LTE	Long-term evolution
LTM	Link trace message
LTR	Link trace reply
MAC	Medium access control
MAC DA	Mac destination address
MAC SA	Mac source address
MCC	Maintenance communication channel

ME	Maintenance entity
MEBH	Mobile Ethernet backhauling
MEF	Metro Ethernet forum
MEG	Maintenance entity group
MEN	Metro Ethernet network
MEP	Meg end point
MIP	Meg intermediate point
MME	Mobile management entity
MP	Measurement point
Mp2mp	Multipoint to multipoint
MPLS	Multi protocol label switching
MTBF	Mean time between failures
MTSO	Mobile transport switching office
MTTR	Mean time to repair
NID	Network interface device
NNI	Network network interface
NTR	Nominal throughput rate
OAM	Operation, administration, and maintenance
ONT	Optical network termination
OTU	Optical transport network
OVC	Operator virtual connection
P2mp	Point to multipoint
P2p	Point to point
PBB	Provider backbone bridge
PCP	Priority code point
PM	Performance monitoring
PON	Passive optical network
PRE	Preamble field
QAM	Quadrature amplitude modulation

QoS	Quality of service
QPSK	Quadrature phase shift keying
RAN BS	Mobile network radio access network base station
RAN NC	Mobile network radio access network network controller
RDI	Remote defect indication
RED	Random early discard
RF	Radio frequency
RTT	Round trip time
SAE	System architecture evolution
SDH	Synchronous digital hierarchy
SFD	Start frame delimiter
S-GW	Serving gateway
SHDSL	Symmetric high-speed digital subscriber line
SLA	Service level assurance
SLO	Service level objectives
SLS	Servuce level specifications
SOAM	Service OAM
SONET	Synchronous optical network
SRC	Source
SRDG	Shared risk domain group
SRG	Shared risk group
SRLG	Shared risk link group
SRNG	Shared risk node group
TCI	VLAN tag control identifier
TCP	Transmission control protocol
TDM	Time division multiplexing
TPID	VLAN tag protocol identifier
UDP	User datagram protocol
UNI	User network interface

UNI-C	Customer UNI
UNI-N	Network UNI
UPS	User plane entity
VDSL	Very-high-bit-rate digital subscriber line
VID	VLAN identifier
VID	VLAN identifier
VLAN	Virtual LAN
WCDMA	Wideband code division multiple access
WRED	Weighted RED
WRR	Weighted round robin

Glossary

The list of definitions has been extracted from MEF 10.2 Ethernet Services Attributes Phase 2. MEF 2009.

All-to-one bundling A UNI attribute in which all CE-VLAN IDs are associated with a single EVC.

Availability performance A measure of the percentage of time that a service is useable.

Broadcast service frame A service frame that has the broadcast destination MAC address.

Bundling A UNI attribute in which more than one CE-VLAN ID can be associated with an EVC.

CBS Committed burst size.

CE Customer edge.

CE-VLAN CoS Customer edge VLAN CoS.

CE-VLAN ID Customer edge VLAN ID.

CE-VLAN ID preservation An EVC attribute in which the CE-VLAN ID of an egress service frame is identical in value to the CE-VLAN ID of the corresponding ingress service frame.

CE-VLAN ID/EVC map An association of CE-VLAN IDs with EVCs at a UNI.

CE-VLAN tag Customer edge VLAN tag.

CIR Committed information rate.

Class of service A set of service frames that have a commitment from the service provider to receive a particular level of performance.

Class of service identifier Information derivable from (a) the EVC to which the service frame is mapped, (b) the combination of the EVC to which the service frame is mapped and a set of one or more CE-VLAN CoS values, (c) the combination of the EVC to which the service frame is mapped and a set of one or more DSCP values, or (d) the combination of the EVC to which the service frame is mapped and a set of one or more tunneled layer 2 control protocols.

Committed burst size CBS is a bandwidth profile parameter. It limits the maximum number of bytes available for a burst of service frames sent at the UNI speed to remain CIR-conformant.

Committed information rate CIR is a bandwidth profile parameter. It defines the average rate in bits per second of service frames up to which the network delivers service frames and meets the performance objectives defined by the CoS service attribute.

CoS Class of service.

Customer edge Equipment on the subscriber side of the UNI.

Customer edge VLAN ID The identifier derivable from the content of a service frame that allows the service frame to be associated with an EVC at the UNI.

Customer edge VLAN tag The IEEE 802.1Q customer VLAN tag in a tagged service frame.

Data service frame A service frame that is unicast, multicast, or broadcast.

EBS Excess burst size.

Egress bandwidth profile A service attribute that specifies the length and arrival time characteristics of egress service frames at the egress UNI.

Egress service frame A service frame sent from the service provider network to the CE.

EIR Excess information rate.

E-LAN service Ethernet LAN service.

E-line service Ethernet line service.

Ethernet LAN service An Ethernet service type distinguished by its use of a multipoint-to-multipoint EVC.

Ethernet line service An Ethernet service type distinguished by its use of a point-to-point EVC.

Ethernet virtual connection An association of two or more UNIs that limits the exchange of service frames to UNIs in the Ethernet virtual connection.

EVC Ethernet virtual connection.

EVC maximum transmission unit size The maximum sized service frame allowed for an EVC.

Excess burst size EBS is a bandwidth profile parameter. It limits the maximum number of bytes available for a burst of service frames sent at the UNI speed to remain EIR-conformant.

Excess information rate EIR is a bandwidth profile parameter. It defines the average rate in bits per second of service frames up to which the network may deliver service frames but without any performance objectives.

FD Frame delay.

FLR Frame loss ratio.

Frame Short for Ethernet frame.

Frame delay The time required to transmit a service frame from ingress UNI to egress UNI.

Frame delay performance A measure of the delays experienced by different service frames belonging to the same CoS instance.

Frame delay range The difference between the frame delay performance values corresponding to two different percentiles.

Frame delay range performance A measure of the extent of delay variability experienced by different service frames belonging to the same CoS instance.

Frame loss ratio performance Frame loss ratio is a measure of the number of lost frames between the ingress UNI and the egress UNI. Frame loss ratio is expressed as a percentage.

Ingress bandwidth profile A characterization of ingress service frame arrival times and lengths at the ingress UNI and a specification of disposition of each service frame based on its level of compliance with the characterization.

Ingress service frame A service frame sent from the CE into the service provider network.

IFDV Interframe delay variation.

Interframe delay variation The difference in delay of two service frames belonging to the same CoS instance.

Interframe delay variation performance A measure of the variation in the delays experienced by different service frames belonging to the same CoS instance.

Layer 2 control protocol service frame A service frame that is used for layer 2 control (e.g., spanning tree protocol).

Maximum number of UNIs The maximum number of UNIs that may be in an EVC.

Multicast service frame A service frame that has a multicast destination MAC address.

Multipoint-to-multipoint EVC An EVC with two or more UNIs. A multipoint-to-multipoint EVC with two UNIs is different from a point-to-point EVC because one or more additional UNIs can be added to it.

Point-to-point EVC An EVC with exactly two UNIs.

Qualified set of service frames The set of frames that comply with specific criteria, such as the arrival time at the Ingress UNI and bandwidth profile compliance, on which a performance attribute is based.

Scheduled downtime A time interval agreed upon by both the subscriber and service provider during which a service may be disabled by the service provider.

Service frame An Ethernet frame transmitted across the UNI toward the service provider or an Ethernet frame transmitted across the UNI toward the subscriber.

Service level agreement The contract between the subscriber and service provider specifying the agreed to service level commitments and related business agreements.

Service level specification The technical specification of the service level being offered by the service provider to the subscriber.

Service multiplexing A UNI service attribute in which the UNI can be in more than one EVC instance.

Service provider The organization providing Ethernet service(s).

SLA Service level agreement or assurance.

SLS Service level specification.

Subscriber The organization purchasing and/or using Ethernet services.

UNI User network interface.

Unicast service frame A service frame that has a unicast destination MAC address.

UNI maximum transmission unit size The maximum sized service frame allowed at the UNI.

User network interface The physical demarcation point between the responsibility of the service provider and the responsibility of the subscriber.

About the Author

Dr. Roman Krzanowski has been working in IT and data networking for over 25 years in Canada and the United States in an industry and research environment. He has been involved in AI algorithm development, large database design, data networking architectures, network metering, and network resiliency, focusing on Ethernet technologies. He holds a Ph.D from the University of London, U.K., as well as several graduate degrees in engineering and philosophy. He was an adjunct professor of computer science in the Polytechnic Institute of New York University. He has 13 U.S. patents granted, has coauthored IETF standard on TWAMP, and has edited MEF standards. He is the author of *The Tao of Network Design.*

Index

Recent Titles in the Artech House Mobile Communications Series

John Walker, Series Editor

3G CDMA2000 Wireless System Engineering, Samuel C. Yang

3G Multimedia Network Services, Accounting, and User Profiles, Freddy Ghys, Marcel Mampaey, Michel Smouts, and Arto Vaaraniemi

802.11 WLANs and IP Networking: Security, QoS, and Mobility, Anand R. Prasad and Neeli R. Prasad

Achieving Interoperability in Critical IT and Communications Systems, Robert I. Desourdis, Peter J. Rosamilia, Christopher P. Jacobson, James E. Sinclair, and James R. McClure

Advances in 3G Enhanced Technologies for Wireless Communications, Jiangzhou Wang and Tung-Sang Ng, editors

Advances in Mobile Information Systems, John Walker, editor

Advances in Mobile Radio Access Networks, Y. Jay Guo

Applied Satellite Navigation Using GPS, GALILEO, and Augmentation Systems, Ramjee Prasad and Marina Ruggieri

Artificial Intelligence in Wireless Communications, Thomas W. Rondeau and Charles W. Bostian

Broadband Wireless Access and Local Network: Mobile WiMax and WiFi, Byeong Gi Lee and Sunghyun Choi

CDMA for Wireless Personal Communications, Ramjee Prasad

CDMA Mobile Radio Design, John B. Groe and Lawrence E. Larson

CDMA RF System Engineering, Samuel C. Yang

CDMA Systems Capacity Engineering, Kiseon Kim and Insoo Koo

CDMA Systems Engineering Handbook, Jhong S. Lee and Leonard E. Miller

Cell Planning for Wireless Communications, Manuel F. Cátedra and Jesús Pérez-Arriaga

Cellular Communications: Worldwide Market Development, Garry A. Garrard

Cellular Mobile Systems Engineering, Saleh Faruque

Cognitive Radio Techniques: Spectrum Sensing, Interference Mitigation, and Localization, Kandeepan Sithamparanathan and Andrea Giorgetti

The Complete Wireless Communications Professional: A Guide for Engineers and Managers, William Webb

Digital Communication Systems Engineering with Software-Defined Radio, Di Pu and Alexander M. Wyglinski

EDGE for Mobile Internet, Emmanuel Seurre, Patrick Savelli, and Pierre-Jean Pietri

Emerging Public Safety Wireless Communication Systems, Robert I. Desourdis, Jr., et al.

The Future of Wireless Communications, William Webb

Geographic Information Systems Demystified, Stephen R. Galati

GPRS for Mobile Internet, Emmanuel Seurre, Patrick Savelli, and Pierre-Jean Pietri

GPRS: Gateway to Third Generation Mobile Networks, Gunnar Heine and Holger Sagkob

GSM and Personal Communications Handbook, Siegmund M. Redl, Matthias K. Weber, and Malcolm W. Oliphant

GSM Networks: Protocols, Terminology, and Implementation, Gunnar Heine

GSM System Engineering, Asha Mehrotra

Handbook of Land-Mobile Radio System Coverage, Garry C. Hess

Handbook of Mobile Radio Networks, Sami Tabbane

High-Speed Wireless ATM and LANs, Benny Bing

Inside Bluetooth Low Energy, Naresh Gupta

Interference Analysis and Reduction for Wireless Systems, Peter Stavroulakis

Introduction to 3G Mobile Communications, Second Edition, Juha Korhonen

Introduction to Communication Systems Simulation, Maurice Schiff

Introduction to Digital Professional Mobile Radio, Hans-Peter A. Ketterling

Introduction to GPS: The Global Positioning System, Ahmed El-Rabbany

An Introduction to GSM, Siegmund M. Redl, Matthias K. Weber, and Malcolm W. Oliphant

Introduction to Mobile Communications Engineering, José M. Hernando and F. Pérez-Fontán

Introduction to Radio Propagation for Fixed and Mobile Communications, John Doble

Introduction to Wireless Local Loop, Broadband and Narrowband, Systems, Second Edition, William Webb

IS-136 TDMA Technology, Economics, and Services, Lawrence Harte, Adrian Smith, and Charles A. Jacobs

Location Management and Routing in Mobile Wireless Networks, Amitava Mukherjee, Somprakash Bandyopadhyay, and Debashis Saha

LTE Air Interface Protocols, Mohammad T. Kawser

Metro Ethernet Services for LTE Backhaul, Roman Krzanowski

Mobile Data Communications Systems, Peter Wong and David Britland

Mobile IP Technology for M-Business, Mark Norris

Mobile Satellite Communications, Shingo Ohmori, Hiromitsu Wakana, and Seiichiro Kawase

Mobile Telecommunications Standards: GSM, UMTS, TETRA, and ERMES, Rudi Bekkers

Mobile-to-Mobile Wireless Channels, Alenka Zajić

Mobile Telecommunications: Standards, Regulation, and Applications, Rudi Bekkers and Jan Smits

Multiantenna Digital Radio Transmission, Massimiliano Martone

Multiantenna Wireless Communications Systems, Sergio Barbarossa

Multi-Gigabit Microwave and Millimeter-Wave Wireless Communications, Jonathan Wells

Multipath Phenomena in Cellular Networks, Nathan Blaunstein and Jørgen Bach Andersen

Multiuser Detection in CDMA Mobile Terminals, Piero Castoldi

OFDMA for Broadband Wireless Access, Slawomir Pietrzyk

Personal Wireless Communication with DECT and PWT, John Phillips and Gerard MacNamee

Practical Wireless Data Modem Design, Jonathon Y. C. Cheah

Prime Codes with Applications to CDMA Optical and Wireless Networks, Guu-Chang Yang and Wing C. Kwong

Quantitative Analysis of Cognitive Radio and Network Performance, Preston Marshall

QoS in Integrated 3G Networks, Robert Lloyd-Evans

Radio Engineering for Wireless Communication and Sensor Applications, Antti V. Räisänen and Arto Lehto

Radio Propagation in Cellular Networks, Nathan Blaunstein

Radio Resource Management for Wireless Networks, Jens Zander and Seong-Lyun Kim

Radiowave Propagation and Antennas for Personal Communications, Third Edition, Kazimierz Siwiak and Yasaman Bahreini

RDS: The Radio Data System, Dietmar Kopitz and Bev Marks

Resource Allocation in Hierarchical Cellular Systems, Lauro Ortigoza-Guerrero and A. Hamid Aghvami

RF and Baseband Techniques for Software-Defined Radio, Peter B. Kenington

RF and Microwave Circuit Design for Wireless Communications, Lawrence E. Larson, editor

Sample Rate Conversion in Software Configurable Radios, Tim Hentschel

Signal Processing Applications in CDMA Communications, Hui Liu

Smart Antenna Engineering, Ahmed El Zooghby

Software Defined Radio for 3G, Paul Burns

Spread Spectrum CDMA Systems for Wireless Communications, Savo G. Glisic and Branka Vucetic

Technologies and Systems for Access and Transport Networks, Jan A. Audestad

Third-Generation and Wideband HF Radio Communications, Eric E. Johnson, Eric Koski, William N. Furman, Mark Jorgenson, and John Nieto

Third Generation Wireless Systems, Volume 1: Post-Shannon Signal Architectures, George M. Calhoun

Traffic Analysis and Design of Wireless IP Networks, Toni Janevski

Transmission Systems Design Handbook for Wireless Networks, Harvey Lehpamer

UMTS and Mobile Computing, Alexander Joseph Huber and Josef Franz Huber

Understanding Cellular Radio, William Webb

Understanding Digital PCS: The TDMA Standard, Cameron Kelly Coursey

Understanding GPS: Principles and Applications, Second Edition, Elliott D. Kaplan and Christopher J. Hegarty, editors

Understanding WAP: Wireless Applications, Devices, and Services, Marcel van der Heijden and Marcus Taylor, editors

Universal Wireless Personal Communications, Ramjee Prasad

WCDMA: Towards IP Mobility and Mobile Internet, Tero Ojanperä and Ramjee Prasad, editors

Wireless Communications in Developing Countries: Cellular and Satellite Systems, Rachael E. Schwartz

Wireless Communications Evolution to 3G and Beyond, Saad Z. Asif

Wireless Intelligent Networking, Gerry Christensen, Paul G. Florack, and Robert Duncan

Wireless LAN Standards and Applications, Asunción Santamaría and Francisco J. López-Hernández, editors

Wireless Sensor and Ad Hoc Networks Under Diversified Network Scenarios, Subir Kumar Sarkar

Wireless Technician's Handbook, Second Edition, Andrew Miceli

For further information on these and other Artech House titles,including previously considered out-of-print books now available through our In-Print-Forever® (IPF®) program, contact:

Artech House
685 Canton Street
Norwood, MA 02062
Phone: 781-769-9750
Fax: 781-769-6334
e-mail: artech@artechhouse.com

Artech House
16 Sussex Street
London SW1V 4RW UK
Phone: +44 (0)20 7596-8750
Fax: +44 (0)20 7630-0166
e-mail: artech-uk@artechhouse.com

Find us on the World Wide Web at: www.artechhouse.com